BUSINESS GEOGRAPHY AND NEW REAL ESTATE MARKET ANALYSIS

Spatial Information Systems

General Editors

M. F. Goodchild
P. A. Burrough
R. A. McDonnell
P. Switzer

BUSINESS GEOGRAPHY AND NEW REAL ESTATE MARKET ANALYSIS

Grant Ian Thrall

UNIVERSITY PRESS
2002

OXFORD

UNIVERSITY PRESS

Oxford   New York
Auckland   Bangkok   Buenos Aires   Cape Town   Chennai
Dar es Salaam   Delhi   Hong Kong   Istanbul   Karachi   Kolkata
Kuala Lumpur   Madrid   Melbourne   Mexico City   Mumbai   Nairobi
São Paulo   Shanghai   Singapore   Taipei   Tokyo   Toronto

and an associated company in Berlin

Published by Oxford University Press, Inc.
198 Madison Avenue, New York, New York 10016

www.oup.com

Oxford is a registered trademark of Oxford University Press, Inc.

Library of Congress Cataloging-in-Publication Data
Thrall, Grant Ian.
   Business geography and new real estate market analysis / Grant Ian Thrall.
      p. cm.—(Spatial information systems)
   Includes bibliographical references and index.
   ISBN 0-19-507636-2
   1. Real estate investment—Evaluation.   2. Commercial geography.   I. Title.   II. Series.
HD1382.5 .T569 2002
332.63'24—dc21   2001021343

9 8 7 6 5 4 3 2 1

Printed in the United States of America
on acid-free paper

Preface

Business geography integrates geographic analysis, reasoning, and technology for the improvement of the business judgmental decision. Without the demonstrated ability to improve the business decision, there is no business geography. This differentiates business geography from the traditional descriptive or explanatory objective of economic and urban geography.

New real estate market analysis in this book is the assembly of information on trade areas, demand, and competitive supply for the purpose of improving the judgmental decision regarding the real estate product. But, real estate market analysis as practiced and taught in universities has generally not conformed to my definition. Rather, real estate market analysis has been a jack-of-all-trades, all things to all people. It has been financial analysis, urban planning, and building construction cost estimation, even law and construction management as practiced by some firms. Each is important to real estate decisions; but, to be proficient, specialists in those areas will certainly agree that each requires very specific and very deep knowledge bases. Today, the depth of knowledge required, and expectation of the marketplace for professional high-level proficiency, precludes one from becoming a master of each and all. Because of the jack-of-all-trades niche that real estate market analysis has occupied, it has been relegated to insignificance and low status in most real estate departments. I believe that limiting real estate market analysis to the focus presented here allows the analyst to acquire a depth of knowledge equal to any academic subject in business or the social sciences, while at the same time being relevant to the real estate decision.

But, there is more to what differentiates new real estate market analysis from what has gone before. Newness also comes from business geography. Business geographers and real estate market analysts have been practicing in parallel for decades. And, academics in both geography and real estate have often pursued the same problems, with the same objectives, but with different methods. Today the methods have converged because of GIS (geographic information systems). The use of business geography technology and methods for the calculation of trade area, demand, supply and absorption analysis, makes for a new real estate market analysis that is more productive, more accurate, and more valuable to the judgmental decision.

David Harvey (1969) wrote in his *Explanation in Geography* that while the emphasis in geography had largely been description, he believed geography could also be an explanatory subject. His book heralded a several-decades-long development of explanatory reasoning in geography. The reader will find philosophy of geography legacy in this book by way of application of descriptive procedures, such as the geographic inventory presented in chapter 7. The reader will also find explanation by way of geographic theory, such as in the ideas of spatial equilibrium and the general theory presented in chapter 3. David Harvey's description and explanation stages of geography were controversial for the time; today they are accepted. But, David Harvey stopped at too early a stage of reasoning. I believe that the value of explanation comes from how it improves decisions.

I believe there to be the following hierarchical categories of geographic reasoning (Thrall, 1995):

1. Description
2. Explanation
3. Prediction
4. Judgment
5. Management and implementation

Explanation gives rise to models that can predict. Prediction is necessary for judgment. The focus of this book is on judgment, structuring information from stages 1 through 3 to improve the real estate decision (stage 4). I believe the above steps are also integral components of real estate as well (Wofford and Thrall, 1997). This book integrates ideas, methods, technologies, and objectives in an opportunistic manner to achieve the goal of providing information to improve the real estate decision.

Geography too has had low status in academia and the professional business community. This is because geography had ignored how its descriptive and explanatory knowledge base could improve decision-making by practitioners. A goal of mine here is to demonstrate that geography adds to the real estate judgmental decision.

In part I of this book, I provide an overview of land use and urban form, its theory, and descriptive analysis, as well as an explanation of market forces and their consequences. This, I believe, is necessary to achieve an understanding of real estate submarkets, and how the various market forces shape real estate submarkets. The real estate practitioner uses and takes advantage of this knowledge in his or her own judgmental decisions.

In part II of this book I proceed through the five basic food groups of real estate: housing, retail, office and industrial, hospitality, and mixed use. Each requires variations from the common theme I develop in part I, and each draws upon a separate knowledge base and technology.

The terms *business geography* and *real estate market analysis* as I use them in this book, are interchangeable, as are the terms *business geographers* and *real estate mar-*

ket analysts. Today, they are interchangeable in industry. Industry is not looking to hire a business geographer, or a real estate market analyst. They are looking to hire someone with the expertise and knowledge base to solve the problem, and get the job done. Another of my goals in writing this book was to bridge the gap between industry and university, and to present a structure and knowledge base that I have found beneficial in getting the job done.

Two people have been particularly important to me in the creation of this book: Dr. Susan E. Thrall and Dr. Larry Wofford. Both have lent their time unselfishly to the creation of this book.

Vail, Colorado   Grant Ian Thrall
July 2001

Contents

PART I

OVERVIEW, THEORY, AND METHODS

I

Introduction

Why Real Estate Market Analysis?

This book shows how to answer questions that provide guidance to the most important decisions in real estate: Should I buy? Should I build? Will the market support the decision? Before developers or investors commit to a project, they must have answers to these questions if they are to remain profitable in the long run.

No generic answers will fit all real estate decisions at all times, because the circumstances differ so much between projects. Therefore, in addition to examples of the methodology that is generally accepted in the industry, this book provides explanations for why the methodology is used. True understanding provides the versatility that is needed in ever-changing markets.

May I build a residential subdivision here? May I build a class A office building there? Such questions are important, but are better dealt with by planners, zoning officials, and attorneys. Is appropriate zoning available? If not, what would be involved in changing the present zoning to accommodate the proposed development? Answering "permission" questions often involves political considerations, which reflect the taste preferences of citizens and politicians and are a byproduct of the local power structure. Therefore, real estate development involves political decisions, as well as market analysis. The market analyst addresses a different scope of issues than the permission questions address.

The market analyst provides information and guidance to the investor, buyer, seller, financier, and planner (See Pyhrr et al. 1989, Vernor 1986, Fanning et al. 1995, Delisle and Sa-Aadu 1994, Brueggeman and Fisher 1996). Funding for development or purchase usually falls upon lending institutions. Lending institutions require market analysis as input for their consideration before they commit to funding. Otherwise, they are jeopardizing the financial viability of their institution. The decision to lend without appropriate market analysis is no more than gambling. Some projects may be successful, some may not. Banking regulations require due diligence in the lending decision to evaluate the exposure to risk by the lending institution. So business geographic real estate market analysis is central to the risk management of those who carry the burden of responsibility for the investment. Market analysis is central to this evaluation of risk. Each project is subjected individually to market analysis to assess the risk of the project. However, the financial institution has many projects it funds. The market analyst assesses the total exposure the financial institution has to risk and thereby must also evaluate those projects collectively. Some financial institutions and real estate investment trusts have adopted strategies of geographic diversification, thereby limiting the exposure to risk within any given geographic submarket.[1]

Market analysis does not end with the decision to build, purchase, or lend. Markets change with time. The market analyst will not foretell every change that will occur in the real estate market; thus, the market analyst must continually monitor the real estate asset to determine how changing local, national, and even global market conditions impact the individual asset and the collection of assets. Results of such monitoring are used over the short term to fine-tune decisions regarding negotiations on contracts and related leasing arrangements. Results of such monitoring are used over the long term to restructure the array of real estate assets of a large corporation to take advantage of new market opportunities and to avoid financial losses.

Real estate assets are becoming inseparable from the successful performance of many businesses that require a physical presence in local markets. The average business has 25 percent of its assets in real estate. Many retail chains have upward of 85 percent of their assets in real estate. The burden falls upon the market analyst to evaluate the geographic distribution of the outlets of a multibranch retail chain, to continually monitor the performance of each of the outlets in the chain, and to evaluate how the composite of all outlets add to the value of the business enterprise. In other words, the market analyst is responsible for providing guidance on the big picture such as how a geographic pattern of outlets could better serve the corporation's objective. The market analyst is responsible for providing insight on the highly focused small picture such as calculating how a new retail outlet will cannibalize revenues from other nearby outlets within the same chain. The market analyst provides guidance ranging from the big picture to the small details, accommodating short-term fluctuations in the market and long-term corporate objectives.

Market analysis is often a requirement of local and regional governmental planning. For example, Florida's Growth Management Act requires that all incorporated governments produce five- and ten-year plans. The plans must include projections for population increase and identification of how the housing needs of the additional population are to be met. Projected housing must be accommodated in local comprehensive plans. The necessary infrastructure must be planned for so that it is implemented in the right location at the right time. The means to pay for the infrastructure must also be decided upon. Placement of infrastructure in the wrong market area—such as on the east side of town when the market is for the town to grow to the west—is wasteful and expensive. The market analyst working for the public sector must accurately project the timing and location of new development. However, when dealing with the public sector, it should be recognized that the political decision is more important to zoning and investment in infrastructure than is impartial market analysis.

The market analyst provides advice on timing: when to sell, when to buy, when to build. The market analyst provides advice in pricing.[2] Too high a price will result in an asset being on the market too long, perhaps missing a peak opportunity. Too low a price will result in a reduction of the return of the owner's investment.

The business geographer performing market analysis is responsible for providing advice at the appropriate scale of geography and relevant location. Real estate decisions are site specific; therefore, market analysis must also be site specific. Analyses at the geographic scale of a nation, region, state, and even county might provide information that sounds interesting, but it may be irrelevant to the decision being made. The relevant market area for the real estate asset must be evaluated and subsequent market analysis performed at the scale of that market area. If a decision is being made to fund a project on the west side of town, the market analysis must be performed at a geographic scale that is accurate and relevant for that location and market area. Recommendations for one submarket might be inappropriate for another nearby submarket. The market analyst is responsible for delivering accurate and timely information for submarkets defined with relevant geographic considerations. The scope of these questions is large, but addressing these large issues is what this book is about.

An inaccurate market analysis will cause a client to miss an opportunity. Or worse, the market analyst could give advice causing the client to suffer a financial loss. Thus, the market analyst must accurately represent the property, and an analysis must not contain errors or omissions important to the client's decision. Market analysis must be inclusive of the relevant information and analysis and must be executed with great care and deliberation.

What Is Real Estate Market Analysis?

Real estate market analysis is the end result of processes designed to investigate and document the myriad factors that determine demand for a particular type of real estate, the supply of competing real estate, and the geographic boundaries of the trade area. Both the demand and supply side of real estate market analysis must be executed at a geographic scale, for a geographic location, and within a time horizon that is pertinent for the real estate asset and is relevant to risk management and the real estate decision maker.

The Demand Side of Real Estate Analysis

Buyers or renters determine their need for a real estate asset. What is needed? How much is needed? Where is it needed? These needs are specific to the individual. The market analyst does not have immediate and perfect insight into the minds of all buyers or renters. Lacking perfect information on all individuals, the market analyst instead looks to the larger market. Instead of calculating what an individual needs, the market analyst calculates what the aggregate market needs in the way of real estate as measured in units such as housing units, square footage, or some other relevant measure.

The market analyst assembles local economic projections, including projections for upturns and downswings. The market analyst then evaluates the competitive position of the real estate product compared to other products in the same geographic market in terms of price, quality, and design. The market analyst's completed report is invaluable input to subsequent risk management evaluation of the proposed project's feasibility, financial viability, and manageability. The risk manager takes the market analyst's report and assembles the information along with reports from other advisors to determine the expected value of the project to the investors; namely, what are the risks and what are the rewards from the *investment*?[3]

Market analysis does have similarities to fortunetelling. Fortunetellers predict future events and how they will affect their client. Likewise, market analysts estimate what the future will be using accepted procedures and explain how that future will affect their client's real estate decision. When an analysis is properly performed, all market analysts should arrive at the same answer. When the answers are not the same, the analysis should be sufficiently explicit and robust to allow the client to determine how, where, and why the analysis differs. The future is not perfectly known. Many events can unfold that might change the future as it impinges on the project. Each location has characteristics that make it somewhat unique. Market analysts must use their experience and judgment to evaluate the importance of phenomena that might make the location unique and thereby an exception to the rule.[4] Therefore, the market analyst presents the results within a range of what is likely to occur under best case, worst case, and mid-range or most likely case scenarios.

Many investors, buyers, and sellers, rely on their intuition for judgmental decisions. How does intuition fit into the world of the professional and skilled market analyst? Intuition must be there at the beginning to set the project in motion. Intuition must support a contract to conduct a rigorous and formal market analysis. Properly performed market analysis can be expensive, and unless it is offset by returns from successful projects, the cost of the analysis can lead to financial loss. Clients whose intuition does not work well for them will soon either avoid market analysis, succumb to an unsuccessful project, or exit the real estate industry.

Intuition about a proposed project must be heeded when the market analysis is completed. The market analyst provides the data to support a decision, but the client makes the decision. Regardless of the sophistication of the analysis, the future is uncertain. Market analyses have a feasible range over which they can successfully enhance judgment. Feasible market analysis depends on data availability, timeliness, accuracy, and application of appropriate analysis, models, and procedures. Availability, accuracy, timeliness, and choice of procedure are interdependent. Occasionally market analysts rely on their judgment to trade one feature for another to achieve the best possible market analysis report, accepting that the ideal analysis is unobtainable in some situations.

Availability

Most data that market analysts use is either purchased from commercial data vendors or obtained from federal, state, or local governments.[5] Some large institutions also use a variety of proprietary data specific to the firm. Commercial data vendors provide information generally only for the top tier market areas. The definition of "top tier" differs by data vendor and ranges from metropolitan areas being among the largest 50 or largest 100. No commercial data vendor provides comprehensive and current real estate data for every market of every size in the United States. Likewise, comparable data may not be available between countries. Internationally, emerging markets are particularly lacking comparable data. The judgmental decision usually requires the same analysis being performed using comparable data between locations, thereby holding everything constant except for location. The analyst can then compare projects at differing locations. However, if the data are not available, then the market analysis must be changed to adapt to the missing data. Often a project without supporting data is not considered. Adapting market analysis often translates into accepting a lower level of accuracy. Not considering a project outside a top-tier market may result in a missed investment opportunity.

Timeliness

The U.S. Census collects population and housing data every ten years. The U.S. Census of Population and Housing contains a warehouse of valuable data for real estate market analysis, including by small geographic area, population count, income, occupation, age information, family composition, average building age, condition of dwelling, and so on. Beginning in 1990, the U.S. Census also began to publish the TIGER/Line geographic data, which include names and locations of streets and address ranges along those streets. Only occasionally does the Census update its housing and population counts outside of the normal census years, and then only for selected market areas using statistical procedures—not actual counts. Only in decennial years does the Census conduct actual counts of population and housing. Likewise, only in decennial years does the Census update its digital street data in the TIGER/Line files. If a subdivision was built after the compilation of the TIGER/Line geographic data, then those streets will not be included in the TIGER/Line geographic data until the next decennial year ending in "0."[6] Availability of timely, accurate Census data therefore depends on what year it is. Private data firms provide data estimates for other years, including TIGER/Line. The accuracy of data from private vendors might also depend on what year it is.

Accuracy

What is the quality of the data available from commercial data vendors? Many factors other than timeliness can affect accuracy. Private data vendors rely on myriad statistical procedures and data other than from the U.S. Census, such as utility hookup data. The procedures and ancillary data used will influence the accuracy of data from private vendors. The price of data ranges widely between commercial data vendors. Population estimates, by county or census tact, projected up to five years or more, vary in price from a few hundred dollars to many tens of thousands of dollars. Paying more does not imply or guarantee greater accuracy.

Analysis

The quality of market analysis depends on how the data are used. Academics and practitioners have proposed a wide range of models and procedures for manipulating data and presenting the results to improve the business decision. The market analyst makes a determination as to which procedure, or variations on a standard procedure, is appropriate for a particular circumstance. The goal of the analyst is to produce a result that can be used to enhance judgment. The analyst proceeds with the belief that circumstances that led to successful projects in the past are repeatable. However, no two projects will have exactly the same set of circumstances. How different from previous experience can a project be before those experiences are no longer valid for future behavior? Market analysis is not separated from the beliefs of the analyst. Market analysts believe in repeatability; however, the environment and society are dynamic. Taste preferences and other important characteristics of real estate consumers may change. Market analysis that takes into consideration only repeatable circumstances might result in information being presented to the client that leads to a wrong decision.

Ultimately, successful clients are themselves well informed about business geography and market analysis and have honed their intuition. The successful client internalizes the market analysis report and uses that knowledge correctly and appropriately to make the right decision.

The Supply Side of Real Estate Analysis

The supply of competing real estate includes that which already exists, that which is in the pipeline and will be subsequently entering the market, and that which can be converted from another existing use to compete in the near future. A first quick look at the supply of real estate may lead one to the erroneous conclusion that determining the supply of real estate is straightforward, two-step process. First, identify the relevant geographic submarket or market area for the particular project. Second, compute the square footage of the type of real estate asset proposed for the project. Each of these two steps have numerous complexities that must be considered, or else the estimation of supply will be in error.

Two components that add to the complexity of computing the square footage of real estate that can compete with the proposed project are conversion and retrofitting of current real estate assets and other real estate projects that are in the pipeline. Each is discussed below, and thereafter, throughout this book.

Conversion, Retrofitting, and the Pipeline

Conversion and retrofitting of real estate assets can make the calculation of supply akin to nailing Jello to a tree! The physical asset of real estate can be upgraded and structurally changed; apartments can be converted into condominiums, for example, and a warehouse can be converted into a retail center. Commercial space can be converted into condominium lofts. A property can be updated by investment in wiring and network interfaces to take advantage of the demand for information technology.

Conversion and retrofitting takes time, and that possibility within the relevant time horizon must be considered when calculating supply.

The pipeline is the supply of a particular type of real estate asset that is in the planning or construction phase and not yet ready for occupancy. Some projects will become public knowledge but forever remain in the planning phase. For political and competitive considerations, secrecy might shield a project from public view and thereby not formally enter the pipeline until building permits are issued. Estimating the increase in the supply of a real estate asset that is in the pipeline can be like a game, where the decision of one decision maker affects the payoff received by the other decision makers. In other words, the individual players are interdependent with one another.[7]

Decisions might be made in secret that will affect the future supply of a particular kind of real estate asset. Without that inside information, the market analyst might advise others regarding the profitability of a particular type of development, and that advice might be in error because supply estimates were in error. As a result, multiple, competing projects might be planned, which in the absence of the others in the pipeline are each evaluated as being profitable. But when competing projects in the pipeline become known and prospective returns are recalculated, information that is more accurate might reveal that the project is unprofitable. It is not uncommon for projects that are highly sensitive to the pipeline to have contracts that specify that up to the time that "the steel comes out of the ground," the developer has the right to stop the project without penalty if another competing project is announced in the same submarket. Examples of such highly sensitive real estate projects would include low-profit–margin grocery stores and multiscreen movie theaters.

Interdependence and the possible deleterious consequences of ignoring interdependent outcomes leads the real estate decision maker to enter the public arena. Instead of keeping the real estate decision confidential, the proposed project might be made widely public with great fanfare in the news media. The project then enters the pipeline to be included as part of the future supply that will be considered by other analysts.

The Geography of Real Estate Analysis

One could say that the motto of real estate is "location, location, location." Geography is the science of location. Therefore, market analysis must conform to established procedures and use contemporary technology of the business geographer.

It is imperative that the market analysis be performed at the geographic scale of the decision being made. All real estate decisions are site specific. Therefore, the market analysis must be performed to reflect that level of geographic specificity. Knowing that the geographic scale and site specificity are important is only a precondition to knowing how to determine what is the relevant geographic scale for a particular project.

The geographic scale for a market analysis of a real estate project depends on the location of those who will receive value from the project in a manner that significantly affects the market value of the project. This is often referred to as the *primary market*. The primary market is calculated as the geographic range over which the most important users of the real estate asset will be drawn. The *secondary market* often envelopes a geographic region beyond the primary market area. The secondary market has a lesser effect on the success of the real estate project and often affects the project indirectly. Still, the secondary market can be quite important for many real estate projects, as it may contribute that additional increment to profit to make the project successful. However, the greater the reliance upon the secondary market, the greater the project's risk. The secondary market is generally more volatile than is the primary market. The reasons for this include (Thrall and McMullin 2000):

- The cost of accessing the real estate project increases with distance to the project.
- Those who would be using and benefiting from the real estate project have an increasing number of other projects to choose from, known as *intervening opportunities*.
- The Internet might substitute for a costly-to-access real estate project.
- The farther a person is from the real estate project, the less likely he or she would be aware that the project exists.

Consider how primary and secondary market areas for restaurants are often calculated. A market analyst's rule of thumb is that 80 percent of restaurant patrons originate within six miles of a typical restaurant. The primary market for a restaurant is then a six-mile radius. Restaurant chains can calculate exactly the range of their primary market area that would inscribe 80 percent of clients. Each primary market area offers a particular demographic composition of prospective patrons. Those prospective patrons originate from their residences, places of work, and places of shopping and recreation. The more people whose demographic characteristics match that of the profile of a patron for the restaurant, the greater the likelihood the restaurant will be successful. The characteristics of the primary market area contribute to the success of a business, and that success is transferred into willingness to pay and ability to pay higher values for locations that offer the desired primary market area characteristics. However, 20 percent of the patrons of a typical restaurant originate outside the primary market's six-mile radius. These people become clients of the restaurant because of some other "traffic generator" that draws customers from a much larger market area.

Thus, the primary market area contributes most to the value that users of the property receive from it. The interactions between a real estate project and the primary market area are less complex; the possibility of unforeseen events is lower than in the secondary market. The secondary market area contributes less to the success of the business, and because of the complexity that comes from a much larger geography, the effects are much more difficult to calculate. However, without the value added from the secondary market area, the real estate project may not be profitable.

The same rule of thumb is used for a national theme park. Eighty percent of patrons might come from a 2,000-mile radius. Twenty percent of patrons might come from the remainder of the globe. In this case, the nation's economy is of great relevance, and the theme park is isolated from the ups and downs of the local neighborhood economy because it does not contribute in large measure to the revenue stream of the theme park.

Simple radial measures, such as those used above, can be useful as first approximations for geographic extent of a market area. But technology allows for greater precision in estimating the primary market area. Few market areas are circular, and most exhibit direction bias. Consider, for example, that there is more trade between the Canadian provinces and bordering U.S. states than there is between adjoining Canadian provinces. The bias in this case is against east–west movement and toward north–south movement. In the United States, the preponderance of neighborhood streets, as well as the national interstate highway system, has been constructed along north–south and east–west axes. Traffic movement follows the transportation infrastructure and therefore is biased against movement against the grain of the 45°, 135°, 225°, 315° angles. Driving time is therefore often a better measure of geographic extent of market area than is mere distance.

Market areas are not only Euclidian in measurement. Physical barriers and governmental boundaries give shape to market areas.[8] Psychological barriers shape market areas as well. There is a geographic bias against interaction between places separated by a high-stress zone. High-stress zones can arise from phenomena such as neighborhoods perceived as having high crime rates and corridors of high-intensity industrial pollution.[9]

Location is of great importance to market analysis, but it is an oversimplification. Methodology is presented elsewhere in this book that can be used to calculate the geographic extent of a market for a real estate project. However, the real estate decision must encompass much more than geography. The real estate decision-maker must integrate location, timing, product, price, and contract into their decision. The real estate decision-maker unabashedly and opportunistically borrows from any discipline that can enhance understanding of how location, timing, product, price, and contract affect real estate.

Components of Risk Management and Decision Making

Real estate decision-makers must integrate the big-five components of real estate into their risk management: location, timing, product, price, and contract. Because real estate projects will vary by the importance of these factors, there must then be many kinds of real estate market analyses, not merely a single simple formula easily programmed into a spreadsheet. An overview follows of how each of the five components relates to real estate risk management, and how these components relate to new real estate market analysis.

Location

All real estate has a location. Geographers measure location either absolutely or relatively. *Absolute spatial location* requires a pair of geographic coordinates, such as longitude or latitude. *Relative spatial location* requires a single distance measurement, such as distance from the city center or distance from the seashore.

The nineteenth-century economist Alfred Marshall coined the terms *site* and *situation* to describe geographic location and its importance to real estate (Marshall 1890, quoted in Thrall 1991). Site is the property itself, referred to as the "pad" in the language of the retail business geographer. The property has width, depth, and shape. It has slope. It has drainage. It has soil of some composition—gravel, clay, sand, and so on. On the site, there might be sewer and water connection. The site has zoning. The characteristics of a site affect the cost of development on that site. Feasibility analysis, namely, determining if a specific project can be successful, must consider the characteristics of a site and the cost of its development.

To the market analyst, the situation of the property is more important than the site.[10] That is, the market analyst is more interested in how the property is geographically positioned with respect to the myriad of phenomena that contribute to the desirability or undesirability of a property. We have already encountered the concept of situation applied in real estate market analysis in the above explanation of primary and secondary market areas. The characteristics of the primary and secondary market areas contribute to defining the situation of a property. Situation addresses what the surrounding uses of land are and how they affect the project. The situation of a site also includes descriptive measurements such as distance to transit, geographically relevant competitive and complementary activities, nearby traffic volume counts, accessibility of travelers to the site, and so on. The market analyst provides this information as input into the larger feasibility analysis.

Product

It is difficult to define a real estate product separate from the site of the parcel. There are legal definitions of the real estate product. Zoning is one form of legal definition. However, zoning normally deals with allowed land use of a neighborhood, not current land use of the property. Some property assessment jurisdictions are required to define property according to its current use. Florida requires all its county property assessors to assign 1 of the 100 official land use codes listed in table 1.1, known as Department of Revenue (DOR) codes. These codes are useful for analysis. For example, they can assist in identifying possible comparable properties and can be useful in making a rough estimation of the supply of comparable properties. However, they are only rough descriptions of the physical property. A theater/auditorium built in the early part of the twentieth century will not be comparable to an end-of-the-century 20-screen theater complex.

Among those general categories market analysts use to classify properties are:

- Price or rental rate of the property
- Vintage or age of the property
- Size of the property and individual units the property might comprise
- Construction quality of the property
- Original intended target niche of the property (high end, low end, etc.)
- Architectural details and style of the property
- Technological capabilities of the property; how it can interface with modern electronic information networks
- Maintenance the property has received and upgrades the property might require
- Contractual restrictions of the property that might enhance or reduce the property's value to prospective users.

The more tightly defined a real estate product is, the closer comparable properties will be that also conform to the narrow definition. However, too narrow a definition allows for error in the estimation of supply of competing product.

Is a newly built residential subdivision with a recreation complex comparable to one with the same characteristics but without a recreation complex? The answer depends on how the prospective renters or buyers value the recreation amenity. If the recreation amenity is a measurable component in the decision of the renters or buyers

Table 1.1. State of Florida Department of Revenue codes for real estate product categories

00 Vacant residential	37 Race track	74 Homes for the aged
01 Single family	38 Golf course	75 Nonprofit service
02 Mobile homes	39 Hotel/motel	76 Mortuary/cemetery
03 Multi-family-(10 or more units)	40 Vacant industrial	77 Club/lodge hall
04 Condominiums	41 Light manufacture	78 Rest home
05 Cooperatives	42 Heavy manufacture	79 Cultural group
06 Retirement homes	43 Lumber yard	80 Undefined—reserved for DOR future use
07 Miscellaneous residential (migrant camps, boarding homes, etc.)	44 Packing plant	81 Military
	45 Cannery/bottler	82 Forest, parks, recreation
	46 Other food processor	83 Public schools
	47 Mineral processing	84 College
	48 Warehouse-storage	85 Hospital
08 Multifamily-(09 or less units)	49 Open storage	86 County
09 Undefined (reserved for use by DOR only)	50 Improved agricultural	87 State
	51 Cropland soil capability Class I	88 Federal
10 Vacant commercial	52 Cropland soil capability Class II	89 Municipal
11 Stores (1 story)	53 Cropland soil capability Class III	90 Leasehold interests (government owned property leased by a non-governmental lessee)
12 Store/office/residential	54 Timberland—site index 90 and above	
13 Department stores	55 Timberland—site index 80 to 89	
14 Supermarket	56 Timberland—site index 70 to 79	
15 Regional shopping	57 Timberland—site index 60 to 69	
16 Community shopping	58 Timberland—site index 50 to 59	91 Utility, gas and electricity, telephone and telegraph, locally assessed railroads, water and sewer service, pipelines, canals, radio/television communication
17 Office buildings (1 story)	59 Timberland not classified by site index to Pines	
18 Office building (multi-story)		
19 Professional buildings		
20 Transit terminals	60 Grazing land soil capability Class I	
21 Restaurant/café		
22 Drive-in restaurant	61 Grazing land soil capability Class II	92 Mining lands, petroleum lands, or gas lands
23 Financial building		
24 Insurance company	62 Grazing land soil capability Class III	
25 Repair service		93 Subsurface rights
26 Service station	63 Grazing land soil capability Class IV	94 Right-of-way, streets, roads, irrigation channel, ditch, etc.
27 Vehicle sale/repair		
28 Parking/mobile home lot	64 Grazing land soil capability Class V	
29 Wholesale outlet		95 Rivers and lakes, submerged lands
30 Florist/greenhouse	65 Grazing land soil capability Class VI	
31 Drive-In/open stadium		96 Sewage disposal, solid waste, borrow pits, drainage reservoirs, waste lands, marsh, sand dunes, swamps
32 Theater/auditorium	66 Orchard groves, citrus, etc.	
33 Nightclub/bar	67 Poultry, bees, tropical fish, rabbits, etc.	
34 Bowling alley		
35 Tourist attraction	68 Dairies, feed lots	
36 Camps	69 Ornamentals, miscellaneous agricultural	97 Outdoor recreational or park land subject to classified use assessment
	70 Vacant institutional	
	71 Churches	
	72 Private schools/daycare	98 Centrally assessed
	73 Private hospitals	
		99 Acreage not zoned agricultural

to locate, then the two complexes are not comparable. The definition should be judged on the basis of being sufficiently narrow while still maintaining a reasonable list of comparable properties. This is one cause of an often-heard complaint levied at developers: All the developments look the same. For the developer, the greater the difference between a proposed project and previous projects, the greater the uncertainty, and the cost of the higher risk is normally greater than benefits received from capturing an unserved market niche. Hence, departure from established norms for successful projects are often limited to higher profit-margin luxury developments.

In the past, the adapt and reuse market in real estate has been very small. However, early in the twenty-first century, this category of real estate among the top-tier markets has become more important. In those markets, a real estate product might be an apartment complex today and a condominium tomorrow. A regional shopping mall might be a collection of retail outlets today and a school or government center tomorrow. A fast-food facility might be converted to a discount car stereo retailer. Nineteenth-century industrial and upper-floor retail property might be converted to a residential loft today.

Real estate products are difficult to compartmentalize. But to conduct a market analysis, real estate products must be placed into categories, so that comparable projects can be identified. This need for market analysts to categorize property is one of the reasons Part II of this book has been arranged following the product categorical approach to real estate analysis.[11] The categories of real estate products that are covered here include residential (for sale and rental housing), retail, office, industrial, hotel, and mixed use. In addition to these categories, other categories exist, and new categories will emerge. For example, assisted living facilities offering limited health care and housing accommodation are an important and growing real estate product category (see Doctrow et al. 1999; Laposa et al. 1997; Mueller and Laposa 1998). The above categories are demarcated according to the use of the property, not according to physical structure or architecture of the property.

The business geographer market analyst might specialize in one of the categories listed above. Specialization allows for an in-depth development of skills and knowledge of markets. Specialization also comes from the value that familiarity with decision makers brings in developing a reputation in a narrow market segment.

Price

Price is a necessary consideration of risk management. Price can refer to the price paid by the owner or the rent or sale price to be realized by the owner. The decision maker might have a specific intended use of the subject property. The business geographer market analyst would forecast expected revenue over time. The greater the price paid for the property, all other things being equal, the less the profit or yield from the investment. Conversely, the lower the price, all other things being equal, the greater the returns.

The assumption "all other things being equal" has a valuable role in general theory, but must be applied with great deliberation because this assumption may not be true. Price is relative. A high price might be justified by strong market potential, and a low price may reflect deteriorating market conditions. Regardless of whether the price is high or low, some price point will exist where above the price point, the proposed project is not competitive with other investment opportunities. Below the price point, the project may be feasible. The financial analyst—not the market analyst—will calculate the critical price point. The market analyst provides information for the determination of such price points, including what the highest yielding project for a particular location would be.

Financial and investment analysts evaluate a specific property as a potential investment. Investment analysis is wide ranging. It includes considerations for the potential use of the property, tax considerations, legal considerations, transaction costs, operating expenses, resale and rental prospects, vacancy, and absorption rates. The financial or investment analyst receives the market analyst's report and considers that information in the context of the market now and at some future time. This leads us to the next component of real estate market analysis, timing.

Timing

The geography of site and situation is continually changing. There is a right time to enter the real estate market with a particular project in mind, and a time beyond which entrance might be ill advised. Evidence that a location has been successful for a particular kind of use in the past is not sufficient to conclude that the location will remain desirable in the future.

Among the geographic considerations that can change is the demographic composition of the population surrounding a site. People tend to age in place. A neighborhood once composed largely of families with children might through time become one composed largely of empty nesters, then the aged, then perhaps young households again. A once successful diaper service in time might find itself without a clientele.

Markets also have cycles. The determinants to cycle and projection of cycles are a sizeable and sophisticated speciality area of market analysis. Geographer Brian J. L. Berry (1991) has written that cycles are naturally occurring phenomena that are highly regular and therefore highly predictable. He believes that there are cycles within cycles within cycles. Very long-term cycles last many decades, and short cycles within them exist perhaps a year.

The real estate product might be a cause of cycles. After new real estate product is built, there is a short time while maintenance costs are low, followed by a longer interval when maintenance costs rise. As maintenance costs rise, decisions are made to reinvest or not. If reinvestment is to be made, at what level should reinvestment occur? Anything short of maintaining and upgrading property to keep it equal to property newly entering the market will lead to a change in the real estate market, perhaps even leading to a change in the use of the property. Market transition might lead to possibly lower yielding uses; this is referred to as *maintenance-based real estate cycles*. A variety of things can hasten, or slow down, maintenance-based real estate cycles.

Developers aim to make their product attractive to their targeted buyers or renters. Attractiveness can be enhanced by adding innovative and trendy design features and thereby bias customers toward their product and away from competitive products. The contemporary design trend of houses with high ceilings and large bathrooms contributes to older houses without those design features being less desirable to prospective purchasers, and thereby having lower resale value. Other design trends are not merely aesthetic; developers of new office buildings must consider the value of making their buildings up-to-date technologically. Older buildings that are not technologically up-to-date might become downgraded even a short time after being constructed. Downgrading can lead to neighborhood transition, known as *filtering*. Lower revenue-generating tenants occupy downgraded office buildings. Filtering within residential developments occurs when households of lower economic standing replace households of higher income.

Cycles Cause Cycles

As a neighborhood ages, it may become unattractive to buyers because of prospective buyers' concern that filtering might occur to successively lower income groups. That concern may itself translate into causing a decline in the value of real estate in that submarket, thereby causing filtering to proceed. Some buyers are therefore biased toward newly built developments. This is particularly prevalent in the United States, where filtering in the second half of the twentieth century had an unmistakable impact on the American neighborhood as it aged.

Planning institutions can in some circumstances counter the cycle-caused-by-cycle effect. For example, in Ontario, Canada, some regional municipalities establish target vacancy rates for specific kinds of real estate product. If the vacancy rate exceeds the target level, building permits will not be issued for more of that real estate product to be constructed. The "time out" will last until the market clears and the target vacancy level is no longer exceeded. Sanibel Island, Florida, and Boulder, Colorado, are two examples of cities in the United States that use variations on this theme. A consequence of such planning is that many individuals and business firms want to locate in those places because they view the planning institution as a desirable amenity. Market analysts must then include cycles, and governmental institutions, in their assessment for discussion (see Goetz and Wofford 1979).

Cycles can lead to investment opportunity. The "malling" of America led to the decline of the downtowns of many U.S. cities. The attraction of shopping in an enclosed, air-conditioned mall, with a great agglomeration of stores, made many central business districts unattractive for retail use. A downward cycle then began because of competition. In many instances, because of filtering, the demographic characteristics of the people that the downtown retailers had previously targeted changed. People with higher socioeconomic profiles moved to the newly created suburbs. Downtowns began a downward cycle; vacancy rates increased and property values decreased. In most U.S. cities of more than 250,000 population, such a downward trend began in the 1950s. Fifty years later, a strong reversal has occurred in the cycle, and the downtowns of most of those same American cities are on the rebound. The cycle thereby brings opportunities to profit, as well as opportunities to lose money.

Some entrepreneurs have invested countercyclically in deteriorating downtown neighborhoods. The cycle allows property to be purchased at a proportion of its peak value. Those entrepreneurs apply their skills to bring about new uses for old properties, including entertainment and the arts, residential lofts in old commercial buildings, and so on.

Terms

Terms must be considered in investment analysis and feasibility analysis; and include contract, payment schedule, risk, and payoff. Office building leases are an example of how contract terms can change the desirability of a project. The monthly rent in isolation of the terms might indicate the project will be successful. However, the new renter might demand, and receive, the first three months of occupancy free, thus decreasing the potential income for the project.

A former renter of retail space might have had the leverage to require a "dark clause" in the contract prohibiting the shopping mall owner from leasing to a new renter in the same category of use as the former renter. So a grocery store that moves to a new strip-mall down the street might leave behind 70,000 square feet of space that is appropriate only for a grocery store, but contractual obligations prohibit the shopping mall owner from leasing to another grocery store. The risk manager for a laundromat/dry cleaner would be advised to review the contractual terms of the anchor store, taking note if there were a dark clause. The laundromat/dry cleaner risk manager would seek to avoid contractual obligations that would prohibit them from closing or require them to continue with high rental payments, if there was a reduction in clientele because of the departure of the traffic-generating anchor tenant.

How Much Market Analysis?

Some developers consider any market analysis as too much. Those developers that survive over the decades might have developed an intuition about what will be successful and what will not. But the landscape is also replete with developers that have had an unsuccessful project, and perhaps not with their own money. The answer then to the question how much analysis depends on the scale of the project and whose money is being used. A major regional shopping mall requires more analysis than a neighborhood strip mall. The regional mall has a larger and more complex market. More time is required to perform the analysis, and there is more money at stake. The answer to the question how much analysis may also be conditional on who is asking.

If it is your money, and a lot of money is at risk, then an appropriate level of market analysis can be used to reduce the risk and provide guidance to the decision being made. The characteristics of whom the analysis is for are also important to the answer. Large corporations are beholden to stockholders. The actions of the managers must be justified. It is difficult for large, stock-held corporations, including real estate investment trusts, to engage in real estate investment and development that is countercyclical to the real estate market. Institutional restrictions then force large corporations to buy on the high side of the real estate cycle, while some upward movement remains. The risk manager should be able to differentiate if the cycle still has an upward trend, of what duration, and of what magnitude. Instead, the countercyclical investor, seldom a large corporation, might buy in as the opportunity arises, expecting the market to bottom out in a reasonable time frame, and proceed to a renewed upward trend. It is also desirable for the risk manager to have access to projections on when the bottom of the cycle will likely occur, and what the duration will be until the next upswing (see Mueller and Wincott 1995). The business geographer market analyst must be knowledgeable about the goals and aspirations of the client, and whatever institutional restrictions may bear on the clients' decisions. Ultimately, the value of any real estate market analysis is measured by how a client's decision was improved because of the analysis.

How much market analysis, at what quality, and at what price, should a developer entering into a new project want? A decision to build a housing subdivision may depend on the timing of the sale of housing lots, thereby generating a cash flow, enabling payment on schedule of debt. A miscalculation of the rate of sale of housing lots may lead to insufficient cash flow and a subsequent downward spiral toward bankruptcy. Real estate market analysis is the price of reducing risk of entering bankruptcy.

An error in judgment to build in a saturated market may result in the real estate asset being considerably diminished in value. Appropriate market analysis can add to the value of the enterprise and should be priced where it yields more benefit than the cost of the analysis. A typical American corporation has about 25 percent of its assets in real estate. Business in retail, particularly multibranch retail chains, have as much as 85 percent of their assets in real estate. Inappropriate and insufficient analysis can lead to diminished value of the real estate asset and therefore diminished value of the corporation. A fast-food retail facility typically costs between $850,000 and $2,500,000 to construct. An office supply facility may cost between $3,000,000 and $5,000,000. An office building may cost from around $12,000,000 to several hundred million dollars. Development cost does not determine market value. Instead, the realized revenue stream generated by the project determines market value, or the amount someone is willing to pay for the real estate asset. A properly executed and properly funded market analysis should enhance the real estate decision, thereby lessening error in judgment by the decision maker.

Concluding Comments

The goal of this chapter was to provide an overview of the interface between business geography and real estate. This overlap is what I refer to as the *new real estate market analysis*. Why is geographic reasoning and geographic technology necessary in the execution of contemporary real estate market analysis? Why is market analysis performed? Who benefits? What are the key components to contemporary market analysis? Real estate market analysis must include analysis of demand, supply, and geography. The decision maker must integrate the five major components to the real estate decision: location, product, price, timing, and contractual terms. Part of the information required for this integration today comes from the business geographer's market analysis. Finally, this chapter addressed the question, how much analysis? The answer to that question depends on who is asking, who is paying, what is at risk, and how much risk is acceptable.

2

Understanding Real Estate Markets and Submarkets

This chapter explains the workings and characteristics of real estate submarkets and the interaction of the submarkets with the larger local market. Market analysts have often defined a real estate market by political divisions such as a county, city, or metropolitan area (e.g., Palm Beach County, the City of Los Angeles, Baltimore–Washington, DC metropolitan area). Market trends are compared and contrasted between real estate markets, such as the growth of Baltimore–Washington, DC versus the decline of Buffalo–Niagara Falls metro areas. Such comparisons at a large geographic scale are valuable because they can identify potential opportunities and potential investment failures. However, as discussed in the previous chapter, the real estate decision is a site-specific decision. Analysis at the appropriate scale and for the appropriate submarket is required to support the real estate decision.

Submarkets are areas within the larger market area that stand out in some important way (see Thrall and Amos 1999; Thrall and McMullin, 2000b). For example,

- Sites within the submarket are at the same stage of a cycle with one another, while perhaps being countercyclical with sites in the larger market or other submarkets.
- Land use within the submarket is homogeneous and differs from land use in other adjacent submarkets (e.g., a submarket of office buildings or retail).
- A submarket might be composed of households of similar demographic characteristics (lifestyle segmentation profiles),[1] and have housing in a similar price range.

The city is an aggregate of submarkets. Each submarket is affected by the whole city; each submarket affects, and in turn is affected by, other nearby submarkets. In other words, submarkets are interdependent with one another and each is interdependent with the whole. Nearby submarkets generally have greater interdependency with one another than they have with submarkets that are more remote. However, some submarkets may be highly interdependent, even though they are distant from one another, such as submarkets of office buildings or industrial park land uses. An urban area will have few office building submarkets that depend on the same geographically large urban market to sustain them; overdevelopment of one office building submarket can have a price effect on another office building submarket. There are generally more housing submarkets than office submarkets in an urban area.[2]

The geographic boundaries of the submarket are defined with an objective of minimizing the variation in the descriptive characteristics and phenomena that characterize the submarket. Usually, the larger the submarket, the more variation there will be within that submarket. Conversely, the smaller the submarket, the less the expected variation among sites. However, if the boundaries of the submarkets are set to be too small geographically, then the sample and population size of sites within the submarket will be too small to produce meaningful statistical measurements, and a true submarket geographic area might be inappropriately split.

The geographic pattern of land use and land values is referred to as *urban form*, and the study of urban form over time is referred to as *urban morphology*. Urban form and urban morphology are a study of those market forces that give rise to existing and changing geographic patterns of land use and land values. This chapter will review the larger market forces that shape our cities and determine real estate values and that result in market trends of submarkets within larger urban areas. Chapter 3 presents three general theories that explain urban land use, urban form, and urban morphology.

Market forces that shape urban form and urban morphology can be grouped into the following nine descriptive qualitative categories:

- Geodemographics
- Transportation
- Interdependence and externalities
- Spatial equilibrium
- Population change and relocation
- Government
- Economic base
- Real estate cycles
- Information and technology.

Together, the above categories of market forces bring about the spatial patterns of urban land use and land values. A change of the parameters within any of these categories will lead to the creation of and changes to urban submarkets.

Each of the categorical market forces are discussed in this chapter. To understand how these categorical market forces affect a city and its geographic submarkets, it is necessary to understand how a business geographer perceives the city.

What Is a City?

Many definitions of cities have been proposed. On the one hand, readers of this book surely know a city when they see one, and, in the final analysis, that might be sufficient. However, to understand the market forces that affect urban submarkets, a more formal definition of a city is required. Indeed, knowing how cities and urban areas are defined sheds light on some limitations to performing market analysis.

Colloquially, the terms *urban* and *city* are often used interchangeably, and I will follow that accepted practice as well. Cities are legally defined in terms of their entitlements, obligations, and geopolitical boundaries. For real estate market analysis, the terms *urban* and *rural* are perhaps more important than the distinction between city and urban. What is urban and what is rural is often defined with clear-cut examples of the polar extremes of urban and rural, while the reality for a specific location might fit somewhere in between.

An urban area does not necessarily coincide with formal city boundaries. The city of Jacksonville in Duval County, Florida, has annexed all the land in the surrounding rural county. Therefore, a large part of the city of Jacksonville's land area is currently rural. There are local expectations that the rural component will quickly become urban; based upon these expectations, all land there is considered urban. So, a place can be considered urban based on expectations and perceptions. Urban land use also often extends well beyond the city limits into unincorporated areas, and might extend to other adjacent cities. These key variables that determine whether a geographic region is considered urban or rural are population density, total population, and population concentration.

The cut-off points are fuzzy for defining what is required for being urban versus being rural. Therefore, on the landscape, the geographic demarcation of what separates and distinguishes urban from rural is also fuzzy. These are not trivial distinctions, as the designations "urban" or "rural" can determine school district policies, law enforcement, public infrastructure, zoning, health and emergency services, and perhaps most important to the business geographer, the availability of data. Availability of data depends on whether the location conforms to an urban or rural designation.

The Census maintains a geographic hierarchy of information that is provided for a place. If the place is defined as urban, there is a greater depth to the geographic hierarchy, and more data are available for each level of the geographic hierarchy. As explained in table 2.1, if a place is urban, the smallest geographic scale for which the Census publishes data is the census block. Census blocks form census tracts. Boundaries for census tracts generally conform to other government boundaries, such as those for cities, counties, and states. So the geographic hierarchy is usually spatially congruent from one level to the next. In contrast, a rural county may have no depth of hierarchy smaller than the county itself. The Census considers an area urban if there are 2,500 persons per square mile, and the place must have more than 50,000 inhabitants.

As illustrated by the Census criteria, a discussion of what is urban and what is not urban can digress into a bewildering array of definitions. Although it is easy to know you are in an urban area when you see one, the formal designation of an urban area can be the difference between having sufficient data to perform the required market analysis or not. Adequate data for real estate market analysis is generally available only for areas designated as urban. Geographic areas that meet the Census Bureau's criteria of urban will have a significantly greater amount of data available from both public and private data vendors than those areas that do not meet those criteria. Even if Census Bureau defines an area as urban, many kinds of valuable data are available only for the top 50 or top 100 urban markets.

Table 2.1. Metropolitan area designations

U.S. Census Bureau reporting units	Description
Metropolitan Statistical Area (MSA)	Before 1983, the MSA was known as a Standard Metropolitan Statistical Area (SMSA). All cities > 50,000 people are designated as an MSA. Statistical measures are reported on the entire MSA, which is inclusive of the entire county that contains the city. So an MSA might include nonurban, rural, areas. The rural area of an MSA is interdependent with the geographic concentration of people in the urban area.
Consolidated Metropolitan Statistical Area (CMSA)	When there are strong interdependencies between MSAs in close geographic proximity, then those MSAs will be combined for statistical measurement to form a CMSA. The majority of the largest MSAs have been combined into CMSAs.
Urbanized area (UA)	The MSA might include rural areas. Rural residents might want the benefits of urban services, such as quick emergency service response, but the low geographic concentration of rural residents makes cost prohibitive the provision of such services. UAs, first designated in the 1950s, remove the rural component from the urban area for statistical measurement reporting reasons. Today, a UA includes: (1) a contiguous place of 2,500 or more people; (2) an incorporated place with more than 2,500 people and at least 100 dwellings in close geographic proximity; (3) adjacent unincorporated areas with at least 1,000 people per square mile; and (4) places that are in between and in close proximity to larger places that satisfy UA designations and that tie the nearby UA places together.
Census tract	Metropolitan areas are divided into smaller geographic units of census tracts for statistical reporting reasons. Tract boundaries are established in cooperation between the Census and the local governing unit. The tract boundaries are established to contain relatively uniform populations sharing similar social and economic characteristics and living conditions. Most tracts have about 4,000 people. When the population of a tract increases from one census year to the next, the tract might be split in the succeeding Census so that each tract has about 4,000 people (see Thrall 1992). The census tract is the smallest geographic area that reports income information.
Census block	The Census block is the smallest geographic area for which the Census reports information. The shape of a census block follows the geographic pattern of the streets, usually a rectangular grid. A block will generally be between two intersections, with residences back to back, not residences on opposing sides of the street. Census block groups are aggregations of census blocks. Often five census blocks compose a block group. A block group is part of the Bureau's geographic hierarchy for reporting geodemographic data. Census block groups when aggregated form a census tract.

Based on presentation of Hartshorn (1992).

Urban Market Forces

The larger market forces that shape cities and determine real estate values were grouped into nine categories earlier in this chapter. The following section introduces and discusses each of these nine categories of market forces.

Geodemographics

Demographic measurements are the wide array of descriptive characteristics of a population, a sample of which is included in table 2.2. The most commonly used demographic measurements by market analysts are population count and a measurement of wealth such as median household income. These measurements are often broken down into subcategories that allow the analyst to hone in on more narrowly defined target groups, such as the number of persons age 45 to 65 years within 7 miles of a location of interest on a map (see table 2.2); the number of persons within a 15-minute drive that have incomes between $55,000 and $75,000; or the number of persons resident within a submarket that have achieved a bachelors degree.

Geodemographic measurements are descriptive characteristics of a population, arranged and ordered by a scale of geography that is meaningful to the analysis. In table 2.2, the geodemographic measurements are arranged by both data attribute and location. Location is measured in the radial distance of seven miles and by radii determined by the distance that can be driven within some time interval. Market analysts use such measurements to evaluate the potential viability of a prospective location for a particular type of use.

Real estate market analysis requires explanation for why the particular kind of land use prevails at that location. With this information in hand, an analyst then wants to predict the answers to questions such as, what will be the geography of the city in five or ten years? If a retail outlet is built here, will it be successful? In the future, where will the population that is expected to be consumers of our products live? Which neighborhoods will be on the decline, and which will be on the rise? Knowing the answers to such questions creates opportunity for the investor, and is the raison d'etre of the market analyst.

Alachua, County, Florida, is used as an illustration in the following discussion on geodemographics. In land area, the county is equally divided between urban and rural land-use patterns. Rural counties, as shown in figure 2.1, surround the county. The population is concentrated in the city of Gainesville, located in the center of the county. Gainesville is the county seat and also the home to 45,000 students attending the University of Florida. The 1999 estimated population of Alachua County was about 195,000; about 66 percent of that number resides within the Gainesville urban area.

Maps and discussion follow on what are generally considered to be the most important geodemographic measurements that contribute to shaping the market forces of most urban areas: population distribution, race, income, and vintage. These four geodemographic measurements significantly contribute to urban form and urban morphology (i.e., the spatial patterns of land use and housing values) and how they change

Table 2.2. Sample geodemographic report

	Population within			
Data field	7 miles	10 minutes	15 minutes	20 minutes
Population (1990)	132,864	91,787	124,142	135,500
Population (1998)	146,703	100,974	136,519	149,816
Population (2003)	154,598	106,033	143,410	158,003
Population density (1998)	1,030.12	1,905.58	1,205.70	766.03
% Population growth (1990–98)	10.4%	10.0%	10.0%	10.6%
% Population growth (1998–2003)	5.4%	5.0%	5.0%	5.5%
Trade Area Size	142.57	53.04	113.35	195.78
Households (1990)	53,499	37,669	50,099	54,159
Households (1998)	60,645	42,684	56,667	61,636
Households (2003)	64,508	45,286	60,122	65,659
Households, married with children (1990)	9,281	5,650	8,453	9,578
Male	72,212	50,114	67,323	73,982
Female	74,491	50,860	69,196	75,834
Race: white	112,608	80,921	105,553	115,640
Race: black	26,710	13,916	23,847	26,719
Race: Asian or Pacific Islander	5,768	4,873	5,566	5,806
Race: other	1,617	1,264	1,553	1,650
Ethnicity: Hispanic	8,672	6,659	8,241	8,762
Age 0–4	9,265	5,503	8,364	9,345
Age 5–9	8,674	5,166	7,846	8,837
Age 10–13	5,679	3,339	5,134	5,803
Age 14–17	4,891	2,921	4,461	5,086
Age 18–24	39,777	33,183	38,251	40,003
Age 25–34	23,111	15,713	21,473	23,543
Age 35–44	20,935	13,178	19,207	21,604
Age 45–54	13,543	8,452	12,383	14,013
Age 55–64	8,150	5,099	7,503	8,448
Age 65–74	6,731	4,283	6,263	7,013
Age 75–84	4,329	2,956	4,088	4,471
Age 85+	1,617	1,182	1,545	1,648
Median age (1998)	28.15	27.11	28.06	28.42
Median household income (1990)	23,452	21,921	23,156	23,847
Median household income (1998)	31,099	29,017	30,673	31,664
Median household income (2003)	37,349	34,446	36,661	38,060
Per capita income (1990)	12,116	11,764	11,995	12,145
Per capita income (1998)	16,910	16,308	16,712	17,071
Per capita income (2003)	19,974	19,072	19,681	20,195
Average household income (1990)	30,089	28,666	29,721	30,385
Average household income (1998)	40,474	38,057	39,787	40,976
Average household income (2003)	47,441	44,142	46,475	48,081
Household income $0–$15K	18,273	14,217	17,344	18,236
Household income $15K–$25K	9,048	6,549	8,538	9,146
Household income $25K–$35K	8,583	5,798	8,023	8,728
Household income $35K–$50K	7,758	5,087	7,221	7,961
Household income $50K–$75K	8,625	5,593	7,967	8,931
Household income $75K–$100K	4,256	2,709	3,848	4,409
Household income $100K–$150K	2,657	1,743	2,410	2,734
Household income $150K+	1,445	988	1,315	1,492
Total housing units (1990)	59,173	41,657	55,370	59,905

Table 2.2. (*continued*)

Data field	Population within			
	7 miles	10 minutes	15 minutes	20 minutes
Occupied units (1990)	53,499	37,669	50,099	54,159
Vacant units (1990)	5,674	3,988	5,270	5,746
Owner occupied units (1990)	24,145	14,617	22,151	25,154
Renter occupied units (1990)	29,354	23,052	27,948	29,006
Gross median rent (1990)	331	337	332	332
Median home value (1990)	71,658	73,456	71,117	71,419
Total consumer expenditures	34,302.22	33,015.36	33,969.15	34,631.06
Total retail expenditures	15,716.87	15,126.10	15,559.28	15,867.52
Occupation: executive/managerial (1990)	8,143	5,565	7,561	8,324
Occupation: professional (1990)	17,011	12,438	15,910	17,067
Occupation: technical (1990)	3,989	2,954	3,794	4,034
Occupation: sales (1990)	7,404	5,333	6,966	7,608
Occupation: clerical (1990)	10,061	6,976	9,451	10,236
Occupation: white collar (1990)	46,607	33,266	43,683	47,269
Total labor force (1990)	63,774	44,381	59,527	64,740
Education: less than 9th grade (1990)	3,245	1,757	2,900	3,324
Education: some high school (1990)	6,213	3,115	5,712	6,703
Education: high school graduates (1990)	12,938	7,228	11,894	13,687
Education: some college (1990)	12,159	7,691	11,296	12,505
Education: associate's degree (1990)	6,923	4,593	6,501	7,171
Education: bachelors degree (1990)	14,538	10,256	13,499	14,673
Education: graduate school (1990)	14,927	11,268	14,073	15,011
Population, age 25+ (1990)	70,937	45,904	65,868	73,067

Table based on Thrall (1999b). (Demographic report for Outback Café situated in Butler Plaza, Archer Road, Gainesville, Florida.)

Urban Florida

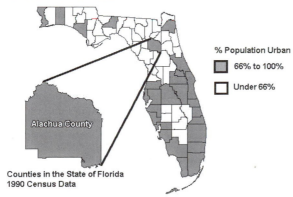

% Population Urban

☐ 66% to 100%

☐ Under 66%

Alachua County

Counties in the State of Florida
1990 Census Data

Figure 2.1 Situation of Alachua County and Gainesville, Florida.

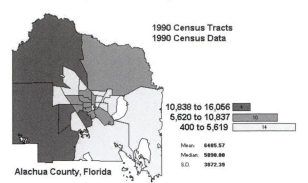

Figure 2.2 Total population of
Alachua County, Florida.

through time. Maps and discussion of each of the four geodemographic measurements follows.

Population Distribution

Figure 2.2 shows the total population of each census tract within Alachua County. The census tracts on the west side of the county are much larger than those in the central part of the county. The map gives the appearance that most of the population of the county lives in those large western tracts, when in fact that is not the case. A more meaningful way to visualize the spatial distribution of population is by way of a map of population density (figure 2.3).

The map in figure 2.3 was constructed by dividing the total population of each census tract by the land area of the census tract, thereby creating the measure of population density. The land area is measured here in square kilometers (a kilometer is about 0.6 of a mile).

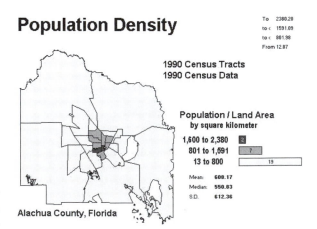

Figure 2.3 Population density
distribution of Alachua County,
Florida.

Notice that the tracts with the highest population density are clustered together at the center of the region, and that population density declines with distance from the urban core (See Clark, 1951; Thrall, 1988). Indeed, this is normally how one or more "urban cores" are defined—namely, where there is a local maximum in population density. Larger cities and geographically more complex cities might have several urban cores around which the population density will decline until meeting the market area of the next urban core. People tend to concentrate around places of desired and frequent access. Most people in Alachua County are employed either at the University of Florida or in government at the nearby administrative buildings. Therefore, the population of Alachua County tends to concentrate around those activities.

Race And Social Areas

Persons with similar characteristics tend to reside near one another. Race is one of those characteristics. The map in figure 2.4 shows the geographic distribution of the African-American population of Alachua County in 1990, by census tract.

The geographic pattern revealed in figure 2.4 and elsewhere conforms to Homer Hoyt's (1939) notion that cities are composed of submarkets, and those submarkets radiate outward from a central core, thereby forming sectors of land use.[3]

Figure 2.5 was constructed using a sample of 33 metropolitan areas and Census Bureau data for the years 1960 and 1970 (Reid 1977). It shows a dramatic change in demographic composition of both urban core and suburban neighborhoods during that ten-year time frame.

A real estate agent engaged in a real estate transaction is required to be blind to such market trends. Instead, a real estate market analyst must interpret how such indicators represent changing submarkets. Therefore, an agent and an analyst have decidedly different obligations and responsibilities.

The U.S. legal system requires that real estate agents, namely, those who are working within the real estate industry and dealing with the buying and selling public, completely disregard race and ethnicity. That is, the real estate agent must be "color blind." The intent is to create a real estate market that is completely impartial and

African Americans

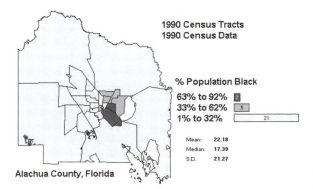

1990 Census Tracts
1990 Census Data

% Population Black
63% to 92%   2
33% to 62%   5
1% to 32%   21

Mean: 22.18
Median: 17.39
S.D.   21.27

Alachua County, Florida

Figure 2.4 African-American population by location.

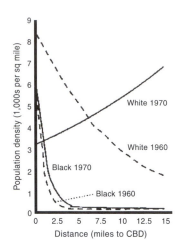

Figure 2.5 Comparative density of two population groups over time (after Reid, 1977).

completely unbiased with regard to race and ethnicity. Because of this, market analysts also often ignore race as a descriptive and predictive factor, regardless of whatever possible relevance race and ethnicity has been demonstrated.

However, it is not a legal requirement for the real estate market analyst to omit race or ethnicity. Indeed, in some circumstances, the federal government requires the real estate market analyst to include race and ethnicity as part of the market analysis. The U.S. Community Reinvestment Act requires that the geographic relationship between bank loans and bank deposits and race be reported.[4] A financial institution will be judged not to be in compliance if their services are not demonstrated to be marketed to minority and poverty-dominated submarkets that are within the larger geographic market area of the financial institution. Also, part of the consideration of where the financial institution's market area is depends on the racial and economic characteristics of the larger region.

Race or ethnicity might be surrogate measures for taste preferences toward or against a particular type of activity or real estate product. For example, say that a chain of assisted living facilities employs a market analyst to determine the most profitable locations for such facilities. The analyst knows based on published market studies that those who use such facilities prefer to reside in facilities not more than six miles from their former residence. Many Asians, like other minorities, have a propensity to geographically cluster together, thereby forming Asian neighborhoods. Also, as a population group, independent of income, Asian households have a bias against placing elderly in assisted living facilities. For an analyst to consider only demographic counts of the total population and not partition that population into meaningful subgroups would be a violation of professional standards and possibly financially ruinous to the company making the investment.

While instances where market analysts use measurements of race can be identified, as in the above examples, in practice, the use of race and ethnicity measurements is the exception and actually seldom used. There are two reasons for this. First, the analyst must ethically and legally maintain professional decorum, and that includes the use of accepted professional methods and practice and not providing advice that effectively "red lines" a neighborhood when that neighborhood would otherwise be receptive to the real estate activity or product. By not including race and ethnicity in the analysis, it is thought that racial or ethnically biased results will not enter into the analysis.

Second, there are better measurements available to the real estate market analyst than race and ethnicity. To the real estate market analyst, race and ethnicity are important only when they are effective surrogate measures of taste preferences and behavioral bias. But, as our society becomes more complex, race and ethnicity have become outmoded descriptive measurements of population subgroups. Instead, the market analyst relies today on lifestyle segmentation profiles (LSP), which are superior at capturing and identifying unique and important traits of population subgroups. Examples will be provided elsewhere in this book on LSPs (see also Brooks 2000; Thrall and McMullin, 2000a; Thrall, et al. 2001).

There are hundreds of descriptive measurements of population subgroups available from the Census Bureau and private data vendors. LSPs arise via geostatistical procedures that allow myriad descriptive measurements to be combined into a small set of commonalities that explain the variation among the population subgroups. Those commonalities are the LSPs. This work dates from the work of geographer Brian J. L. Berry during the 1960s (Berry 1964, 1969, 1970). Berry later was a consultant in the creation of the first commercially available LSP (see Thrall and McMullin 2000a), which has become commonly used in marketing and real estate market analysis. Applications of LSP to real estate market analysis are presented elsewhere in this book. Many LSP terms have entered our active vocabularies, such as "dinks" (dual income, no-children [kids] households), and yuppies (young urban professionals). The use of LSPs yields better results for the real estate market analyst than do crude, one-dimensional measures of race. An African-American yuppie and a European American yuppie might have everything in common except for their skin color. Race then becomes an ineffective explanatory measurement for taste preferences and behavioral bias by a demographic subpopulation, while the LSP is more effective at predicting the market. The real estate market analyst would find figure 2.5 more useful if, instead of race, densities of people by LSP were graphed.

The real estate market analyst has the task of evaluating how a market composed of many different kinds of people will respond to a particular real estate product at a specific location. This is a decidedly different task and objective than an evaluation of how individual businesses or categories of business deal with demographic subgroups, as in the case of the Community Reinvestment Act.

Income

The map in figure 2.6 reveals submarkets by household income. The high-income sector in the example of figure 2.6 radiates outward toward the northwest. The high-income sector of Gainesville from the nineteenth century through the early 1950s was instead in the northeast. Those neighborhoods can still be seen forming a small, localized northeast submarket. Beginning in the 1950s, there was rapid growth of the University of Florida on the west side of the city on land that was formerly agricultural. Also, an interstate freeway was developed on the west side of town. Together, these events reshaped Gainesville's urban form and urban morphology. For example, there was a cessation of expansion, and even decline, of the northeast high-income residential submarket, and instead a large, high-income residential submarket was created to the northwest.

Median Family Income

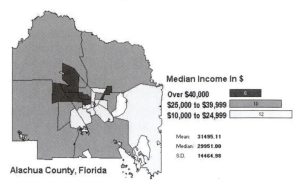

Median Income In $

Over $40,000

$25,000 to $39,999

$10,000 to $24,999

Mean: 31495.11
Median: 29951.00
S.D. 14464.98

Alachua County, Florida

Figure 2.6 Income distribution in Alachua County, Florida.

Several universal themes are depicted in figure 2.6. First, households with similar incomes tend to agglomerate. Urban form is not random, but instead shows high spatial regularity and is highly predictable. Without such regularity, there could be no real estate market analysis. Second, households with like characteristics are spatially distributed within sectors radiating outward from the central urban core. Cities have sectorial patterns of land use, and geodemographic patterns of population that conform to those sectors. Sectors of households of similar incomes arise in large measure because, through time, the trend has been for households to move outward as shown in figure 2.5. Also, as population subgroups increase in numbers, new development occurs that is adjacent to the prevailing geographic concentration of that population subgroup. The adjacent new development usually extends the radial pattern outward from the market core.

Therefore, the income distribution contributes to the geographic shape of the city. Knowing where those income groups are located is important to developers of every kind of real estate. Retail such as Talbot's is targeted toward higher income women; Talbot's strategic management must consider the location of those women in their decisions on where to construct new stores. Similarly, strategic management of Target must consider as part of their location decisions the geographic patterns of the income groups that patronize their retail stores.

The geodemographic ring-study and drive time analysis of table 2.2 shows the breakdown around a central point of population by income (see Thrall 1999b). A homebuilder or retail developer or office building developer must be aware of where the households are that they want to target with their products and ensure that their product is geographically positioned to accommodate and attract that target population. From another standpoint, someone who owns land at a given location must realistically evaluate the demographic composition around that location, including drive-time demographic studies, so that a realistic development profile of the property can be created.

Vintage and Time

Geodemographics includes descriptive measurements of the population, as well as descriptive measurements of development by location over time. Age of development

Median Year House Built

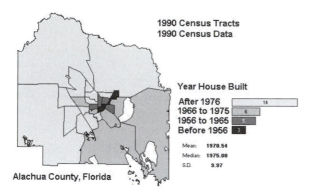

1990 Census Tracts
1990 Census Data

Year House Built

After 1976   14
1966 to 1975   6
1956 to 1965   5
Before 1956   3

Mean: 1970.54
Median: 1975.00
S.D.   9.97

Alachua County, Florida

Figure 2.7 Vintage of a city.

is one of the more important descriptive space-time measurements about the urban built environment.

In figure 2.7, houses built before a half-century ago are generally concentrated at the central urban core. The urban core is composed of neighborhoods with the oldest vintage. The neighborhoods with the youngest vintage are at the periphery of the central urban market. In general, as shown in figure 2.7, the age of the dwelling decreases with increasing distance from the urban core. This often-repeated geographic pattern gives rise to the analogy that cities grow like rings on a tree. Cities begin their development at a central core and then expand outward. In the case of Alachua County, the central core was a regionally important nineteenth-century railway station. This circumstance, and the supporting commerce, was the reason that people lived near what would become downtown Gainesville. Through time, as the city aged, new development was added adjacent to, but farther outward, from the previously existing development. Also, more places that people want to frequently access, referred to as *nodes*, have been added at the periphery. Those newly introduced nodes, including shopping and employment centers, become the stimulus for the subsequent round of urban development.

The map of urban vintage in figure 2.7 shows a space-time pattern of development. The real estate market analyst needs to consider the space-time trajectory of development so that land can be acquired at the right location, and sufficiently ahead of the space-time path of development, so that the necessary land might be purchased at significantly lower cost. At the same time, purchase of land too far in advance of the market support for its developability will increase its holding cost and therefore the total cost of the development.

In the map of figure 2.7, new urban development is predominantly occurring in the previously nonurban areas to the north and west of the older urban core. If development were to occur in already established areas, then that development would be referred to as *urban infill*. Market analysis of urban infill development must take into consideration possible spillovers, both positive and negative, of existing nearby development, including submarket real estate cycles.

Figure 2.8 combines the concepts of population density (figure 2.5) and that of vintage or time (figure 2.7) (Anderson 1985; cited in Yeates 1998). Figure 2.8 allows

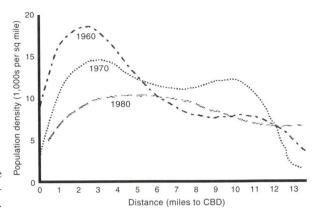

Figure 2.8 Change in space-time density for Detroit (after Anderson 1985, cited in Yeates 1997).

for a quick first approximation to the future of population density by location, extrapolated from past trends. A retail store located two miles from the central business district in figure 2.8 in 1960 might not have sufficient demographic support for remaining profitable in 1980.

Transportation

In the United States and elsewhere, the most important characteristic for establishing what the density of population will be as land has been developed is the predominant mode of transportation prevailing at the time of development. Mode of transportation has a dramatic effect on the population density and on the willingness of the population to move away from their favored nodes.

In the nineteenth century and earlier, the predominant mode of urban transportation was foot travel. Foot travel; is expensive in terms of time and effort. To minimize distances between origins and destinations, such as dwellings and places of work, people clustered together near their place of work, thereby forming high population density. Successive transportation eras have each given more geographic freedom to those able to afford the transportation. In the United States we have gone from the foot, to the trolley car, to the automobile, to the freeway in less than a century. Most mid-sized and larger American cities today are having their urban land-use patterns shaped by the freeway.

Some people today express a dislike for the low-density, auto-dependent urban environment that has come about since the 1950s. They look nostalgically back to earlier vintage urban landscapes. In response to what is seen as a significant niche real estate market, some new urban developments have been designed with the characteristics of urban form of earlier vintage cities. Walt Disney World's Celebration in Orlando, Florida, is an example of a mixed-use development with land-use patterns and density that hearken back to earlier urban vintages.

From the post–World War II era to the start of the twenty-first century, the nominal cost of transportation, as measured by the price of gasoline, has increased by about five times. Over the same time, the median income of American households increased in nominal terms about ten times. In other words, the real cost of transportation declined during the second half of the twentieth century. The post–World War II era was characterized as being the most rapid and expansive of suburban development. Therefore, if transportation cost is the most important factor in allowing or enticing people to locate farther from places of desired or required frequent access, then people are not expected to soon give up their suburban lifestyles.

However, the market analyst must consider geographic movement and location in a broader context than mere transportation cost measured by the price for a gallon of gasoline. Instead of transportation cost, we generally measure accessibility. Within the context of the city, accessibility is more a function of time required to travel than cost of travel.

The once-barren freeways of the 1950s are today congested with fellow travelers. For the middle-income household, travel is more a consideration of willingness to allocate time than how much it costs in dollar terms to get there.[5] Because people generally value their nonwork time equal to the value of their work time, then an extra half hour commuting each way to work can be expensive and perhaps a compelling motivation to locate near those places of desired and frequent access, including work, shopping, recreation, school, worship, and so on.

While the middle- and high-income households have experienced a dramatic decline in the real dollar cost of transportation over the past 50 years, the same cannot be said for low-income households. With incomes fixed at the bottom of the income distribution, a location decision by a low-income household is an effective means to increase disposable, after-transportation-expenditure income. If a low-income household chooses a private mode of transportation, then the lower that household's income, the greater the percentage of income allocated to transportation expenditures. This is in large part because of the high fixed cost of transportation: the purchase price of the vehicle and insurance. The amount of transportation allocation can easily exceed one-third of a low-income household's income. Therefore, it is more compelling for the low-income household to locate near those places of desired or required access than it is for a middle-income or high-income household.

If the low-income household shifts to a public mode of transportation, it can dramatically lower the percentage of income allocated to transportation. However, this constrains low-income households to locate in places with superior access to public transportation. Those places are often at the city center or other important urban node where public transportation lines merge. Those locations are also where land prices per square foot are the greatest. Paradoxically, low-income households are often willing to pay more for central locations than higher-income households because of savings in transportation cost. Higher-income households have a greater range of locational opportunities, and that greater supply of acceptable locations generally results in a lower price per square foot for land.

Low-income households face a dilemma today about where to reside. The changing geography of the city has created this dilemma. Jobs since the 1950s have been moving to the suburbs. Those suburbs are generally not serviced by public transportation. Retail and the service sectors have followed the middle- and higher-income households to the suburbs. Figure 2.5 demonstrates the locational trend by race. A graph like that of figure 2.5, but for households divided into the categories of low-income and middle income and above would result in a similar pattern for most large cities. In other words, the density of higher income households has increased with increasing distance from the city center, while the density of higher income households has declined at the city center. Lower income households—black and white—have remained in the urban cores.

Members of low-income households cannot search for a job or keep a job because they cannot afford the requisite private transportation required to access employment opportunities. With low job skills, wages may not cover the necessary private transportation expenditures. The changing geography of the city has resulted in public transportation increasingly not linking low-income households to desired destinations, particularly the desired suburban employment opportunities. The requirement for the low-income household to locate near public transportation junctions is diminished. Therefore, the benefits to the low-income household of residing at the urban core have decreased. Paradoxically, the diminished attraction that central locations now have to lower-income households might be a contributing market factor in the gentrification of those same urban core neighborhoods. Why should a low-income household pay a premium in rent to reside downtown if by doing so transportation costs are not reduced and access is not increased?

Early in the twenty-first century, low-income households are becoming geographically trapped without access to job opportunities. This geographic trap for low-income households provides a setting that fosters high rates of crime against nearby people and property. Affordable suburban housing with affordable access to employment and shopping is seldom within the reach of low-income households.

For middle-income and higher-income households, access is viewed in terms of time for getting from origin to destination, rather than in terms of transportation cost. For example, ask someone from any of the top-tier cities how far away a place is? The response normally given will be in time: "that is about a 30-minute drive from here, except during rush hour, when it will be 60 minutes to 90 minutes from here." Time measurements of distance are often more important than cost measurements of distance. When the decision maker is translating distances into time for their decisions, such as to go to the shopping center or not, then the market analysis should also be performed in terms of relative location as time as well. Therefore, among market analysts, transportation is considered primarily in terms of time, not monetary costs. Monetary transportation considerations are generally not considered in real estate development decisions.

To measure transportation access, market analysts have generally performed geo-demographic ring studies around the property being evaluated. Table 2.2 includes an example of a ring study report.

A ring study reports the demographic characteristics within a specified radial distance from a point. Figure 2.9 shows circles around a point in Gainesville, Florida, the same point as that used to derive the geodemographic report shown in table 2.2. Distance bands for one, three, and seven miles are shown. Also shown are major roads including state highways and Interstate 75.

However, as discussed above, more relevant to those making decisions than distance measured in terms of radial miles is access measured in terms of time. Where to shop or where to buy a home is based on proximity to shopping, schools, and so on, where proximity is measured in terms of time. Therefore, measurements of proximity based on drive times are preferred to rings based on radial distances. Figure 2.10 is a ten-

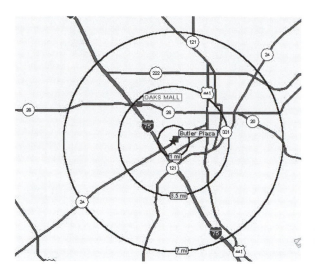

Figure 2.9 Radial distance bands around a central location (after Thrall 1999).

minute drive-time map from the same location table 2.2 is based on. Notice how the transportation network—namely, the location of the major roadways—determines what can be accessed within the allotted time. Boundaries on drive times are seldom circular.

Interdependence and Externalities

The single word that best describes a city is interdependence. Without interdependencies, cities would be just a random occurrence. Instead, the benefits of interdepen-

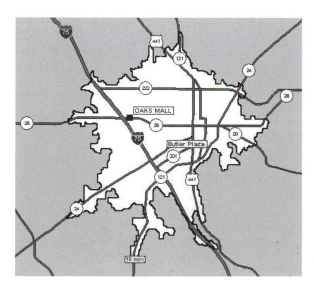

Figure 2.10 Driving-time distance band (from Thrall 1999).

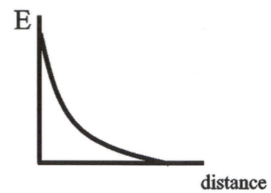

Figure 2.11 Externality density function versus distance from source.

dence cause people and their activities to assemble in close geographic proximity. The greater the geographic proximity, the greater the impact one individual, or group of individuals, will have on others.

One of the conditions that must hold for a *perfectly competitive market* to exist is for there to be independence between all the actors participating in the market. In other words, the actions of one person cannot and do not affect any others. However, the city exists because of interdependence, and the city creates interdependencies. A city, and the market land values and land uses within the city, does not conform to strict principles of perfect competition. Examples of interdependence include:

- The negative effect that one poorly maintained house can have on resale values of other houses in the neighborhood
- The positive effect that an attractive recreational facility such as a golf course or park can have on nearby properties
- The positive effect that access to desired or necessary destinations can have on housing values.

The force, either positive or negative, which emanates from the many interdependencies and contributes to land prices or land uses, is referred to as an *externality*. We measure the intensity of the externality across distance from its source, as shown in figure 2.11. Externality intensity is often greatest at its its source and declines with distance away from the source. The source may be mobile or stationary and may exist at a point or along a line.

Externalities can be good, or bad, positive or negative, relevant or irrelevant, as well as substitute or complements to land (Thrall 1987). Externalities can affect individual and collective well-being. Examples of negative externalities that affect housing land use include noise from traffic along a highway, congestion on a highway, smog, and crime against people and against property which may originate from a neighborhood.

The definition and intensity of a negative externality can differ between population subgroups. Noise might arise because of high traffic count. High traffic count to a household would be considered a negative externality. At the same time, a retail shop owner would consider high traffic count to be a positive externality. Examples of positive externalities that affect housing land use include benefits obtained from a beautiful park, an exciting museum, good urban design and urban planning that increases the quality of the life in the urban-built and natural environments, and proximity to persons of similar income and taste preferences, thereby providing for the threshold for the provision of goods and services that one enjoys and desires to consume.

Externalities affect all categories of land use. On the one hand, households might want to locate near places of employment such as manufacturing. However, proximity to manufacturing facilities can provide a nuisance and hazardous environment for the household, resulting in restrictive regulations on manufacturing that otherwise would not be imposed. So high proximity of households to manufacturing facilities can be a negative externality to both manufacturer and household. A retail location might benefit from externalities that arise because of proximity to an interstate highway with high traffic volume. Because retailers are willing to pay a premium for access to potential customers traveling the interstate, commercial property values that receive those externalities can be higher than commercial property elsewhere.

Taste preferences and even location itself can be a determinant as to what an externality is and the intensity of that externality. Geography alone can cause the intensity of the perceived benefit from a nearby park to range from being considered as highly desirable to indifferent. Consider a household located in the high population density submarket of an urban downtown. Those households' yard may consist of a few square feet of balcony. The introduction of a park nearby will substantially increase their well-being because they otherwise lack open space. Going from a position of scarcity of a desirable good to availability of that good will greatly increase the households' well-being. Because the desirability of residing at the urban downtown location has increased, property values will increase. Consider the same household residing in the distant low-density suburbs, perhaps on several acres of land. The otherwise identical suburban household does not have a scarcity of open space by virtue of the low population density. If a park is built near them that is in every way the same as the urban downtown park, the suburban household's well-being is not expected to increase by as much as their downtown counterpart, if at all. Indeed, the park might be perceived as a nuisance to the suburban household. Therefore, relative location can change an externality effect (Thrall 1987).

Externalities create problems and opportunities. A city is a balance between positive and negative externalities. When positive externalities outweigh negative externalities, a submarket can grow (see Grissom et al. 1991). As a collection of submarkets grow, the city grows. Alternatively, when the negative outweighs the positive, urban decline sets in.

Spatial Equilibrium

When economic phenomena are analyzed, the analysis is normally carried out in terms of price–quantity relationships: how much suppliers are willing to offer on the market at a given price, and how much the market is willing to purchase at a given price. Such an analysis requires price–quantity equilibrium. The intersection of supply and demand, as shown in figure 2.12, establishes the market equilibrium price. At this

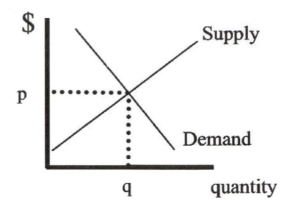

Figure 2.12 Market-clearing price.

price, consumers as a group are willing to buy the same amount producers, as a group, are willing to sell. The equilibrium price is often referred to as the *market-clearing price*.

Land is a consumer good that shares some characteristics with normal consumer goods, such as those that would be analyzed using figure 2.12. Land can be bought on time. It provides benefit to the consumer. However, many characteristics of land are not shared with normal consumer goods. Land has location measured in an absolute sense with latitudinal and longitudinal coordinates and measured relatively as distance to someplace else. The absolute location of land cannot change. The relative location of land can change, but as a result of the decisions of others that change utilization of land around the site. These unique characteristics of land make it highly susceptible to externalities, as discussed in the preceding section. Figure 2.12 includes no locational component, nor does it include any form of locational equilibrium. Therefore, figure 2.12 cannot capture the full "location, location, location" component of real estate.

Figure 2.12 offers equilibrium conditions for the analysis of price-quantity relationships. However, the real estate market analyst must consider locational relationships, in addition to price-quantity relationships. Because location matters, geography matters, and we must therefore bring geographic equilibrium into real estate market analysis.

To understand geographic or spatial equilibrium, consider those market forces that pull households away from places that they frequently access versus pushing them closer. C. C. Colby first wrote about these concepts of spatial equilibrium in his pioneering 1933 publication in the *Annals of the Association of American Geographers* (Colby, 1933). Population density and land values generally decline as access to desired destinations declines. So why doesn't everybody move farther out and receive the benefits of lower land prices and larger neighborhood lot size? Because by moving farther away from the places of desired access, greater transportation expenditures are incurred and more important, more time must be allocated for commuting. By moving toward those places that are frequently accessed, transportation time and costs are reduced but with accompanying disadvantage of higher land costs. Colby reasoned that geographic or spatial equilibrium occurs when there is a balance of centripetal and centrifugal forces, namely the forces to move away are balanced by the forces to move closer in. The market is then in spatial equilibrium when households have no incentive to relocate.

Spatial equilibrium is much broader than simply moving toward or away from some node. Spatial equilibrium market forces explain the trajectory of movement of land values at a particular location as a result of a change in market forces. If the land values are below spatial equilibrium, then households will move in, bringing the price of land up until successive households are no longer willing to make the move and pay the price required for that location. Land values adjust. Land prices are the driving mechanism that will make successive households indifferent to alternative locations. For application of spatial equilibrium to commercial and industrial land uses see Thrall (1991), and Kolbe and Evans (2001).

Say some phenomena, quantifiable or not, results in a higher level of well-being received by a household from residing at a location. As households become generally aware of the higher level of well-being available at that location, they will attempt to move there. Their voting with their feet, and pocketbooks, results in an increase in land prices at the desired destination. Conversely, say that some phenomena diminishes the level of well-being at a location; as households subsequently avoid that location, land prices for housing there become diminished.

Prices increase or decrease by that amount sufficient to make households indifferent between moving to that location or away from that location. In economics, equilibrium is achieved when there is no incentive to increase or decrease what has become the market-clearing price. In geography, spatial equilibrium is achieved when there is no incentive to move to a location or to move from a location (see Casetti 1971).

Opportunity arises when the urban area and its submarkets are not in spatial equilibrium. Land prices move in the direction required to bring the system into spatial equilibrium. The real estate market analyst should be able to identify those submarkets that are out of spatial equilibrium, and should include in the report their evaluation of the general land price trajectory that will come about as the submarket moves toward spatial equilibrium.

In figure 2.13, household welfare, $u[s]$, is measured on the vertical axis, and distance from a node is measured on the horizontal axis. Welfare at location s_1, $u[s_1]$, is shown to be higher than what may be obtained on average elsewhere—namely, U. Any quantifiable and nonquantifiable phenomena could cause this situation to occur. Households that move to location s_1 cause the price of land to rise there. At location s_2, household welfare, $u[s_2]$, is lower than the spatial equilibrium, U. Households leave s_2, thereby leading to a decline in land values at s_2. Household welfare at s_2 increases as housing/land becomes more affordable with the decline in land values there.

When the urban area and its submarkets are in spatial equilibrium, the price of land has adjusted up or down to offset any localized higher or lower benefits. To capitalize on opportunity, the real estate market analyst must be aware of the phenomena that can change the spatial equilibrium of a submarket. The real estate market analyst ideally knows about these changes before the average member of the market knows, and thereby can intuit the trajectory of land values. The real estate market analyst therefore engages in information arbitrage.

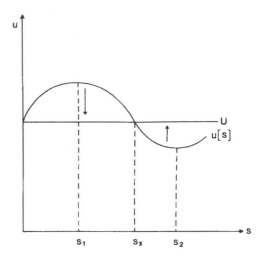

Figure 2.13 Spatial welfare surface
(Thrall 1987).

Population Change and Relocation

Spatial equilibrium is both a cause and an effect of population movement. There are spatial equilibrium forces at work at the global, national, state, city, and submarket levels. Migration is one component of population change.

Population change comes about by natural increase and by migration. *Natural increase* is the differential between birth rates and death rates (see figure 2.14). Demographic transition is the process of society being transformed from having high birth rates and high death rates, stage I, to that of low birth rates and low death rates, as shown in stage IV.

The industrial revolution began in England and Wales around 1740. The industrial revolution brought with it an improvement of the human condition, particularly a decline in infant mortality. In stage II, birth rates remain at traditional levels while death rates decline; therefore, population increased.

In stage III of figure 2.14, birth rates begin a general decline, while death rates continue to decline. Birth rates decline for a variety of reasons. Anthropologists and sociologists would refer to stage II high birth rate as *cultural lag,* a remnant of an earlier era when the likelihood of a child surviving to adulthood is low. To have two or three children survive to adulthood, a household might need to have seven or more children. The death of a child during the life of a parent was expected and normal in stage I. Early in stage II, with more and more children surviving to adulthood, the wealth of the family as a traditional economic unit would increase.

Sometime during stage II, households internalize that a greater number of children are surviving to adulthood, and a premium is no longer placed on having so many children. The wealth of the family, when divided between so many individuals, might decline. As a result, households have fewer children, and the birth rate declines, as shown in stage III. Stage IV is reached when the transition is complete, and birth rates and death rates are once again nearly equal.

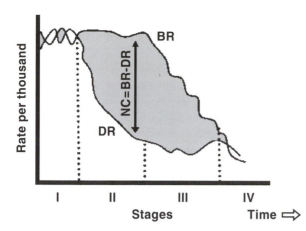

Figure 2.14 Demographic transition.

The rate of birth is inseparable from culture and cultural norms. Today, among many less developed nations, children are an integral part of the wealth creation process of the household: More children mean more wealth for the family unit. In industrialized nations, labor laws generally prohibit children from contributing to the economic wealth of the household. Instead, the choice to have children, and the number of children to have, is tied more to the decision to consume and a budgetary matter. At the start of the twenty-first century, the cost of raising a middle-income child from infancy, not including college, is more than $250,000, which is more than the cost of an average single-family dwelling. So the greatest expenses a family will decide upon in rank order are to have one or more children, then housing, and then transportation.

From about 8000 BCE, to 1 CE, the global human population grew to about 250 million. The earth's population at the beginning of the twenty-first century is more than 6 billion and doubles about every 15 years. Some consider this rate of population growth to have been a great environmental disaster. The very survival of our species is thought by some to require a check on further population increase of this magnitude. Another viewpoint is that humans have proven that they have the ability to advance technology to maintain, and even improve, the quality of human well-being and human experience, even with such rapidly increasing numbers.

The real estate market analyst has a more limited role and responsibility: to determine the rate of population increase and how that increase will affect real estate markets. Globally, particularly among less developed nations, natural increase continues to drive real estate markets, even as those economies are in steady state or even in decline. In the United States, the natural increase of population has had, and continues to have, a dramatic affect on real estate markets. Consider the evidence of the effect that the baby boomer generation has had on the U.S. housing market as these boomers have proceeded through the various stages of their life cycles.

A *cohort group* is a subset of the total population that shares a significant characteristic, such as age. The Census Bureau and private data vendors, as shown in table 2.2, publish data on cohort groups generally in five-year age increments. Market an-

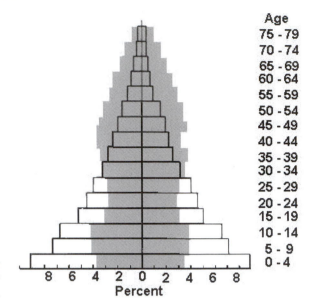

Age
75 - 79
70 - 74
65 - 69
60 - 64
55 - 59
50 - 54
45 - 49
40 - 44
35 - 39
30 - 34
25 - 29
20 - 24
15 - 19
10 - 14
5 - 9
0 - 4

8 6 4 2 0 2 4 6 8
Percent

Figure 2.15 Cohort groups (after Berry, Conkling, and Ray 1987).

alysts might need to combine several cohort groups to reflect the stage of life cycle, as this has a bearing on the demand for the particular real estate product.

For comparison, two classic age forms are shown in figure 2.15 (Berry, Conkling and Ray 1987; see also Demeny 1977, Freedman and Berelson 1977). Male cohorts occupy the right side of the axis and females the left. Each bar is proportional to the number of persons occupying the age-by-sex cohort group. Vertically adjacent bars can be added to form larger cohort groups as required for a specific analysis. The unshaded pyramid of figure 2.15 is common for most developing countries. The United States also exhibited the pyramid form through the early part of the twentieth century. The advent of the baby-boom era brought on the cohort group form as shown in the shaded portion of the figure. At the beginning of the twenty-first century, the leading edge of the baby-boom generation is turning 50 years of age. The bulge in the shadowed portion of figure 2.15 begins at about age 75, so this diagram, although for a northwestern European country, is roughly what the U.S. age distribution by cohort group will appear like in the year 2025. People require different forms of housing according to their age. Figure 2.15 reveals the demand for housing by cohort group. Submarkets might reveal different cohort age distributions than the urban market or the nation.

U.S. housing prices began to soar in the early 1970s in part because the leading edge of the baby-boom generation was at the stage of household formation when they were purchasing their first homes. The bulge in cohort group brought about a rapid increase in demand for housing, and consequently a rapid increase in housing prices. This rapid increase in housing prices brought with it a change in housing type within the United States.

Condominiums have been commonplace in many countries, and in some submarkets of the United States, including ski resorts. However, condominiums had not been

Change In Population By State

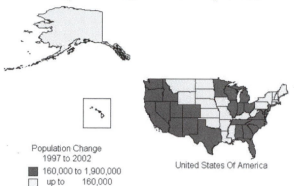

Population Change
1997 to 2002
■ 160,000 to 1,900,000
□ up to   160,000

United States Of America

Figure 2.16 Change in U.S.
population by state.

as accepted as had the single-family dwelling in most housing markets. Indeed, in many housing markets before the 1970s, there was no body of laws in place to regulate or allow condominiums. However, because condominiums are more affordable to first-time homebuyers than single-family dwellings, condominiums have become an increasingly frequent type of housing purchase.

Population can also change because of in-migration and out-migration. The attraction of California to the baby-boom generation led those cohort groups to leave regions that offered lower levels of attainable welfare such as Ohio, Manitoba, and Arkansas. Thus, housing prices continued to increase in California after they had become stabilized elsewhere. In other words, as in the earlier discussion of spatial equilibrium (figure 2.13) the perceived attainable welfare levels of California (real or not) were like region s_1, and that for Arkansas was s_2. People moved to California. In-migration, the baby-boom cohort bubble, and an economy that was able to provide employment to home purchasers resulted in soaring housing prices. Population density increased. Congestion increased. The level of attainable and perceived welfare in California then became lower than alternative destinations, including Colorado, Nevada, Oregon, Washington, Arizona, and Florida. The city of Las Vegas swelled since 1985 to change it from a small desert oasis to a city of more than 1.25 million people, mainly ex-Californians. At the beginning of the twenty-first century, California is losing 500,000 persons each year to other states, primarily Colorado, Nevada, Oregon, Washington, Arizona, and Florida. The demographic profile of an out-migrant from California is college educated, around 35 years of age, earning the U.S. median income, and largely white, but blacks and other racial groups with similar LSP characteristics are leaving as well. Most were born in California. Those out-migrants are seeking higher levels of welfare attainable elsewhere, so they are moving to other national submarkets. However, as shown in figure 2.16, California's population will continue to increase, primarily due to in-migration from Latin America and Asia. Although the ex-Californian might achieve a higher level of welfare in Colorado, an immigrant from Vietnam might achieve a higher level of welfare in California than he or she had in Asia.

California's in-migrants tend to be younger, of lower economic status, and with less education than the out-migrants. For California's new in-migrants, the expected

Older Population Distribution

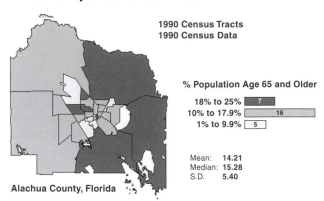

1990 Census Tracts
1990 Census Data

% Population Age 65 and Older

18% to 25%   7
10% to 17.9%   16
1% to 9.9%   5

Mean: 14.21
Median: 15.28
S.D.   5.40

Alachua County, Florida

Figure 2.17 Population distribution age 65 and older.

level of attainable welfare for them and their families is well above that from where they originated. If California's economy improves, incomes there will rise, thereby increasing the level of attainable welfare and diminishing the rate of out-migration to other states. Also, as congestion and spoiling of the natural environment occurs in migration-destination states, their attractiveness also diminishes. Thus, the centripetal and centrifugal forces will one day be back in balance.

A real estate market analyst must consider the demographic trends of natural increase, in-migration and out-migration, and age distribution. More important than the count of population by cohort group for the nation is the count of population by age for the various submarkets being analyzed.

Figure 2.17 shows a map of the percentage of the population by census tract age 65 and older. Figure 2.18 shows a map of the percentage of the population by census tract within the age range of kindergarten through grade twelve, or age five through seventeen. By combining adjacent cohort groups, the school-age cohort group was

School Age Population Distribution

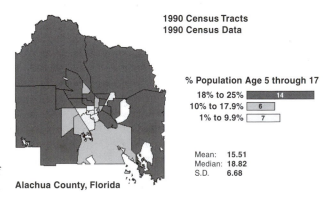

1990 Census Tracts
1990 Census Data

% Population Age 5 through 17

18% to 25%   14
10% to 17.9%   6
1% to 9.9%   7

Mean: 15.51
Median: 18.82
S.D.   6.68

Figure 2.18 Distribution of school-age population.

Alachua County, Florida

derived. The maps in figures 2.17 and 2.18 demonstrate that different submarkets exist for population by cohort group. The western census tracts have a larger-percentage of school age population and a lower percentage of the population aged 65 or more. The converse is true of the eastern portion of the country.

The type of housing required differs as people proceed through their life cycles. A map like that shown in figure 2.17 would be helpful in deciding where to construct an assisted living facility because when deciding where to build such facilities, the elderly prefer to remain near the neighborhood that they had been living in. A youthful population with little equity and no children might require low-price condominiums or rental accommodations. A large population of children, might indicate a market for traditional housing with several bedrooms.

Government

Local governments have considerable impact on the city and therefore on the recommendations of the analyst (see Wolf 1981, Feagin and Parker 1990). Real estate development is both a market consideration and a political decision. The following discussion deals with issues related to development and the local government. This discussion includes the entitlement process of zoning and permitting and how local government decisions affect real estate values. The general role of government in real estate is first addressed.

What is the role of government in real estate? The answer is a subset of the broader question on the interactive roles of geography and government (see Bennett 1980). An answer to the question of the role of government in real estate development and market analysis arises out of distinguishing between public goods and private goods. Table 2.3 illustrates one categorical approach to distinguishing between public goods and private goods.

In table 2.3, a good is considered *excludable* if someone can be kept from consuming it. This is a requirement for a private good because, without it, there is no way to extract a price from the consumer. Without excludability, there is no mechanism for requiring payment; without payment, the good will not be provided.

A good is *rival* if, by consuming the good, others are prohibited from receiving benefit from the consumption of that item. *Pure private goods* are both excludable and rival. Examples include a vitamin pill or a cup of coffee.

Pure public goods are neither excludable nor rival. Pure public goods will not be provided without government involvement. At the geographical scale of the city, police

Table 2.3. Categorical definition of public versus private goods

	Rival	Nonrival
Excludable	Pure private	Quasi-public
Nonexcludable	Quasi-public	Pure public

Source: Musgrave (1959)

protection falls into the category of a pure public good. If the police department is successful in keeping the overall crime rate down, then all within the city benefit. Environmental regulations and their enforcement also fall into this category.

Goods falling outside the pure public and pure private cells of table 2.3 are the most problematic. Those goods share characteristics of both public goods and private goods. Local urban streets are an example of a good that is rival, but not excludable. Their consumption by large numbers of people makes them rival. Congestion at rush hour is evidence of such rivalry.

An often-repeated political phrase, "run government like a business," ignores the nature of the goods provided by the respective private and public sectors as defined in table 2.3. For a business to exist, it must provide only those goods that have excludability, and thereby obtain revenue. But if goods that do not have excludability as a characteristic are to be provided, then a government must supply those goods. A legitimate debate, however, is what should be the role of the private and public sectors regarding the provision of goods and services that are not purely public and not purely private. This is determined by the value system of the society and how meritorious the society considers their provision. Local governments provide bundles of goods—private, public, and impure—in response to the value system of the society. Charles Tiebout (1956) hypothesized that the more closely the provision of goods provided by the local government conforms to the wishes of the local population, the higher will be the demand by that population to reside within the geographic boundaries of the government and thereby have access to those goods. This may result in higher land values within that government jurisdiction than otherwise would prevail. In other words, people vote with their feet; people move to governmental jurisdictions whose provision of goods conforms most closely to their own preferences, subject to budgetary and other locational considerations. Table 2.4 provides a small selection of illustrative activities important to real estate and a short explanation for the role of government in these activities.

Instead of four categories of goods as in table 2.3, one alternative approach is to array goods and services along a continuum, ranging from pure public to pure private (Buchanan 1965). Starting from the polar position of pure private goods, some goods at that end of the spectrum may have some degree of publicness, for example, a room in a house might be shared by two siblings. The siblings might effectively have a "room club." Members of the household share the kitchen, kitchen utensils, a television set, and so on. The household members effectively have a household club, as the sharing of certain goods is expedient and cost effective. Residents of an urban submarket might join to form a church, temple, or synagogue, golf club, or tennis club, each of which is a club but on a larger scale than the household club. Residents of several urban submarkets might form a city—a larger club still, as are states and nations. This "theory of clubs" then provides a sense for how organizations come into being to provide goods and services that have degrees of excludability and degrees of rivalry and might provide various levels of externalities because of interdependencies.

Housing subdivisions and industrial parks conform to this theory of clubs. Developers of housing subdivisions might restrict new houses to be consistent with architectural design standards such as minimum square footage. An industrial park might be restricted by its developers to compatible types of industry with the objective of minimizing negative externalities between tenants.

Table 2.4. Government and real estate market analysis

Example item	Explanation	Justification	Amount of subsidy or expense
Education	Quasi-public good providing meritorious externalities to society	Without public subsidy, the quantity of education provided by the private sector would be lower than acceptable to society.	Complete subsidy available at the lower levels of elementary and secondary school. Partial to no subsidies at higher levels such as college
Sewers	Quasi-public good providing negative externalities to society in the absence of their provision	Without governmental requirements, the quantity of waste removal would be lower than acceptable to the larger society.	Part of the debate in every local government is how much should be paid for this infrastructure out of general revenue funds versus other means such as impact fees. New development can be excluded from connecting to sewer lines; therefore, a price for connection can be extracted.
Streets	Quasi-public good because of ease of access makes for the good being non-excludable; however, because of congestion there can be rivalry in consumption	Contemporary transportation technology requires roads, and without roads a place cannot economically develop.	Like sewers, should local roads be paid by taxes levied by government, or other means such as impact fees on new development?
Economic development initiatives	Quasi-public good providing externalities to society as a whole, but more to some individuals than to others	Without government involvement, coordinated local action is unlikely because of inability to exclude from benefits.	Most economic development organizations are public–private partnerships, with each sharing in the costs.
Zoning and permitting	Zoning restricts land use to be other than what the normal operating market would otherwise bring about	A city exists because of interdependencies, and because of these interdependencies there is market failure; namely, a departure from perfectly competitive market forces. These interdependencies create externalities, both positive and negative. Zoning is intended to minimize the effect that negative externalities have that result from nearby incompatible land uses, and maximize the effect that positive externalities have in order to increase the general well-being of the population.	Zoning and permitting can range from (1) "police action" where a reasonable but not necessarily the highest use of land is left for the owner while other uses are prohibited, (2) purchase of some of the rights and entitlements of the property such as "urban development rights"; (3) purchase of all the rights and entitlements of the property, thereby transferring ownership of the property to the government.

(continued)

Table 2.4. (*continued*)

Example item	Explanation	Justification	Amount of subsidy or expense
Environmental regulation	Environmental regulations limits the use of land in some manner other than what might be considered best by the landowner or land user.	Without government limitations, because of interdependencies both at the moment and in the future, land might be used in a manner that deleteriously affects persons nearby, including their ability to receive the full benefits of their land. Environmental regulations are also justified on the basis of their merit.	Same as zoning or permitting
Taxation policies	Governments have an obligation to provide a package of public goods and services, and other meritorious goods and services, and must pay for those activities through either taxation or possibly inflation.	State and local governments must maintain a balanced budget. State and local governments provide many of the public goods that directly affect individuals. Those governments generally use property, sales, and to a lesser extent income taxes to generate their revenues. Federal government relies primarily on income taxes. For reasons based on merit the market has been biased toward house ownership by way of mortgage interest deductions.	While the U.S. has a mortgage interest deduction, Canada does not.

Local government has the most direct impact on real estate development. The "club" of local government is a mirror of the taste preferences and values of the population that reside there, subject to constraint of the larger club of the U.S. legal system.

Land as a good is more difficult to place in one of the categories into table 2.3 than, say, pencils or aircraft carriers. This is because land is not just one thing, but possesses many attributes. The court system recognizes this and accommodates legal issues dealing with land being divisible into its attributes. Write all the distinguishable rights of land onto slips of paper, and place those paper slips into a glass jar. Ownership of a parcel of land is the equivalent of ownership of the bundle of rights and obligations within the jar. Unlike other goods, each of those slips of paper can be individually sold or otherwise given over to another individual or a government, and thereby taken out of that jar, leaving ownership to be defined by the remaining slips of paper.

Because of externalities, local governments normally require the developer to seek approval before any land-use change. Change in land use without authorization is a component of land ownership that has been removed from overall land rights in most local government jurisdictions. The landowner can sell their development rights of the parcel to government or a conservancy organization. The parcel will not then be developable, but the owner can retain the right of current land use, and even sell that right to another or leave it to heirs. Changes in zoning can change the bundle of rights and thereby change the range of opportunities available for the land parcel. Environmental regulations are similar to zoning.

Who is to pay for goods provided by government? Henry George (1879) a nineteenth-century journalist, also asked that question. In our terminology, Henry George reasoned that those who choose to reside in a city do so because of positive externalities that result from interdependencies. Those externalities serve to increase land prices. So the benefits of residing in the city spill over to land. Henry George reasoned that the greater the benefit, the greater the land value. Because city government has the responsibility to provide an array of goods and services like those mentioned in table 2.4, and because those goods and services must be paid for, Henry George reasoned that the property tax is the appropriate tax for local governments.

Regardless of the structural form of taxes, the analyst must consider the extent to which a tax creates bias toward or away from real estate, and if a tax increases or decreases the affordability of real estate. For example, in the United States, mortgage interest deductions effectively increase the disposable income of middle-income households who purchase housing via a mortgage. That biases housing consumption toward private ownership, and in most instances increases the ability to afford more expensive housing. In Canada, which does not have a mortgage interest deduction, the bias is toward rental housing and lower affordability (see table 2.4). However, other benefits can be argued in support of the Canadian scenario (see Bourassa and Grigsby 2000, Weicher 2000; also see Thrall 1987, 116–129).

A precondition for development is consideration of the availability of infrastructure that will support new development. Some local jurisdictions require this as part of development review. Will the new project increase traffic volume enough to require widening the road to maintain current levels of infrastructure quality? This is a task that differentiates the real estate market analyst from, say, an urban planner. The urban planner evaluates the impact a development will have on the urban area, whereas the business geographer evaluates whether there is a market to support the development at that location and if that location is the best for the development.

Economic Base

The most important market force that drives the local real estate market is that market's economic base.[6] As households earn income, a percentage of that income is allocated to savings and also to purchases of a variety of consumer goods and services. Persons who are paid to provide those varieties of goods and services in turn allocate a percentage of their income to other goods and services, and so forth. The number of times the dollar circulates is referred to as the *multiplier effect*. There are two approaches to deriving the multiplier effect. The economic base multiplier approach is followed here. There is also a related but slightly more complex approach known as input–output multipliers.

The *multiplier* refers to the amount of increase in income, jobs, and spending that is beyond that created by an initial investment. Real estate is a crucial link in the multiplier process because housing is such a large percentage of a typical household's budget. Also, some form of real estate is generally required in the provision of goods and services.

The local economy is divided into two parts: The *basic sector* refers to goods and services produced within an urban area, but either sold outside the urban area or sold to those whose incomes are not derived locally. The *nonbasic sector* refers to goods and services produced and sold within the urban area itself. The fundamental belief behind such analysis is that a "place cannot survive by taking in its own laundry." In other words, the existence and growth of the urban area depends on goods and services it produces and sells beyond its boarders. Also, the size of the nonbasic sector is thought to depend on the size of the basic sector.

Say that total income in a city is $400 million, and income from the basic sector (BA) is $100 million. Income from the nonbasic sector (NBA) must then be $300 million. Therefore, for every $1 spent in BA, there will be an additional $3 generated in NBA. If $50 million in new BA is somehow created, then NBA will increase by 3 times that amount, or by $150 million. The multiplier effect results in an increase in total local activity by $200 million ($150 million + $50 million). So the total local economy, because of the $50 million in new BA, has grown to $600 million.

Local chambers of commerce generally calculate and make available the local multiplier; of course, there is an incentive to report inflated measures. In general, the larger the city, the greater will be the multiplier because there is more opportunity for dollars to be spent locally. Dollars not spent locally are referred to as *leakage*. The smaller the city, the greater the expected leakage.

There are many problems with the basic–nonbasic approaches. One is the forced separation between BA and NBA, and the relatively greater importance given to the BA in urban growth. Other problems are in the calculation of BA and NBA. In practice, employment is often used rather than money flows because data on money flows are generally not available. This approach also generally ignores wealth effects of the type, say, that stimulate retirement communities and college towns. For example, a person might have saved for retirement while working in New Jersey and then moves to a golf course community in Florida. That person is consuming housing and other goods and services in Florida based on job activity in an earlier time period and in another geographic area, which is not included in basic and nonbasic calculations of the Florida community. The same problem arises for college towns like Gainesville, Florida, where students receive money from student loans and parents and spend that money in Gainesville, thereby providing jobs. But, the students' initial dollar spent was neither created locally nor spent in the BA. So there is a problem in distinguishing between what is basic and what is nonbasic.

Real estate is dependent on both BA and NBA. As BA and NBA rise, demand for real estate products and services increase. The business geographer reports on the expected growth of BA and NBA. Later in this book, in the product-specific chapters such as on office markets (chapter 6), more will be discussed on how BA, NBA, and breakdowns of both BA and NBA into smaller economic units can be used in the calculation of real estate product-specific demand.

Real Estate Cycles

Urban growth and development ultimately depend on development of the nation as a whole, including both urban and rural sectors. Cities grow as Gross Domestic Product (GDP) increases; cities decline if the GDP has a long-term decline. The production of GDP must take place somewhere. The amount of GDP produced at a particular location can have multiplier effects in that community, as discussed above, thereby stimulating further wealth and more GDP. Part of that increase and flow of GDP goes into real estate.

The increase in GDP is not smooth or continuous. Indeed, GDP can decrease. Overall, the GDP has had an upward trend in the United States since the foundation of the nation. However, there have been a series of downturns in GDP. Fluctuations in GDP can be quite high, leading to rapid urban development as well as downward spirals.

Because cycles have such an enormous effect on real estate, both positive and negative, the study of real estate cycles and methods for their prediction has created a complex, specialized subfield (see Pyhrr et al. 1990, Born and Pyhrr 1994, Pyhrr et al. 1999). A general understanding of cycles, their causes, and interpretation is important to the real estate market analyst. The calculation of cycles is best left to specialists in the subfield.

At a general level, there are two types of cycles: those induced by endogenous market forces, and those induced by exogenous market forces. Endogenous forces arise from inside the system such as inventory buildup. Business then slows down production until inventory clears. The slow down in production results in higher unemployment. Due to the multiplier effect, such downturns can lead to a downturn in the local urban economy, including real estate. These cycles are normally of three to four years in duration. Investment cycles are longer term, about seven to eleven years in duration. Based on projected rates of profit, business might invest in capital to expand production. That investment has local multiplier effects, leading to economic upswings. Once that investment has been made, it will be amortized over a reasonable period, thereby resulting in a decrease in multiplier stimulus.

Analysts are now called upon to compare and contrast the position that a place may be within a cycle. That information is then used to guide the geographic diversification and product diversification of an investor's or financial institution's portfolio. (For discussion and issues related to geographic diversification, including substitution of product diversification for geographic diversification, see Cheng and Black, 1998, Grissom et al. 1991, Hartzell et al. 1987, Eppli and Laposa 1997, Mueller 1993, Mueller and Ziering 1992, Rabianski and Cheng, 1997.) Because there is uncertainty in how cycles will ultimately behave over time, the risk manager will internalize this information in his or her decisions to minimize risk of loss, while taking advantage of localized rise in real estate values.

Exogenous cycles result from outside forces. Politicians might wait until eighteen months before their reelection to provide some stimulus to the local economy, through, for example, the construction of a new school or navy shipyard. As politicians have cycles of reelection, then so too may their constituency. Immigration is seldom smooth. A nation's economy might suffer a downturn for political reasons, resulting in a wave of immigration, resulting in an increase in demand for housing in those destination markets. In the five years before Hong Kong was repatriated to China, Hong Kong dollars flooded into the North American west coast housing markets, inserting Hong Kong capital into a low-risk haven. Geographer Brian J. L. Berry (1991 p. 10) has documented that "within the inherently high noise levels of history, prices and economic growth move in synchronized rhythms . . . in approximately 55 year waves within which 25–30 year cycles are embedded." The long waves of 40–60 years duration are known as Kondratieff (1935) cycles, named after an early proponent of the idea. Perhaps at the root of these long-term cycles are cycles of innovation and technology change. Businesses take advantage of the opportunity to increase productivity by investment in new, more productive equipment. But if the technology is rapidly changing, the business that adopts last can have a productivity advantage over early-adopting competition. Thus, periods of rapid technology innovation might lead businesses to disinvest and then later to invest, thereby causing cycles that lag behind technology change.

Real estate cycles create opportunities for acquisition of wealth. In real estate, as in the stock market, you make money by buying low and selling high. In real estate, though, the investment period is often measured in multiyear intervals versus the minute-by-minute fluctuations taken advantage of by the stock market day-trader. The analyst specializing in cycle analysis will provide advice on when to buy in, when to hold, and when to sell, and on the duration of the peaks and valleys of the cycle. Considerations of cycles are particularly problematic for large firms. To buy low and sell high in real estate translates into being a countercyclic investor. Many large firms cannot justify to their stockholders an increase in real estate portfolio of derelict properties based on the expectation that some day they will be worth considerably more. Successful countercyclic investors are usually individuals who are investing their own funds.

Maintenance-Based Cycles Revisited

Real estate cycles can be highly localized, with their cause and effect being limited to a single neighborhood. One such important real estate cycle was introduced in chapter 1 and referred to there as maintenance-based cycles. Maintenance-based real estate cycles are analogous to endogenous investment cycles discussed above. However, the decisions made are by homeowners, and the collective interdependent actions of homeowners affect the viability of many nearby businesses, thereby setting off their own business cycles. This form of cycle is important because it explains the growth, maturation, and decline of many residential neighborhoods in the United States.

Housing in the United States is today constructed with materials with an expectation that the housing capital will last about 30 years. Separate from the land, for a 30-year-old dwelling to be equivalent in all respects to a newly built dwelling, the additional expenditure on the physical structure would, in the limit, be equal to the cost of replacement.

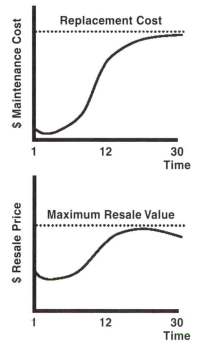

Figure 2.19 Housing life cycle.

The top graph in figure 2.19 illustrates a generalized S-shaped maintenance cost curve, asymptotically approaching the replacement cost. The bottom graph illustrates a generalized price curve for resale value of dwellings in a neighborhood. There is a low and flat maintenance period of about seven to ten years, then maintenance costs begin to rise.

Maintenance is a choice, referred to as a *variable cost*, VC. The *fixed cost*, FC, is the purchase price of the dwelling. Once purchased, that original purchase price cannot be changed. The total cost, TC, can be expressed as

$$TC = FC + VC \tag{2.1}$$

Within a small standard deviation, housing usually sells at a price that is the average amount per square foot for the neighborhood. If the average price is $80 per square foot, and the subject property has 2,000 square feet of conditioned space, then the first estimate of the sale price of the house would be $160,000. For that reason, households are generally advised not to overinvest and not to build beyond their neighborhood. In other words, the more the household spends on VC beyond the average expenditure within the neighborhood, the smaller will be the percentage of those additional expenditures the household is likely to recover upon selling the house.[7]

There are two groups of actors with payoff matrix described in table 2.5. Group A and group B actors are numerous and geographically mixed together in a local submarket. Say the base value, or base rent, of both groups of dwellings is the same

Table 2.5. Prisoner's Dilemma applied to maintenance-based real estate cycles

| | Actions of group B | | | |
| | Renovate | | Not renovate | |
Actions of group A	Group A	Group B	Group A	Group B
Renovate	100	100		
	−10	−10	100	
Not renovate	+7	+7	−10	
	+2	+2	+7	100
	+2	+2	+2	+2
	+___	+___	+___	+___
	101	101	99	102
		100		
		−10		
	100	+7		
	+2	+2		
	+___	+___		
	102	99	100	100

at $100. The lower right pair of cells has the $100, and neither group maintain or renovate their property. The important part of figure 2.19 is that costs of maintenance of a property increase as the property ages, and if maintenance expenditures are not made, then the property will deteriorate. In one scenario, group A chooses to renovate or maintain their property, while group B does not (upper right pair of cells in table 2.5). Each group A player spends $10 for renovation. Because the dwelling itself is in better condition, A players can increase their rent by an additional $7. Because there are many A players, there is a neighborhood price increase of $2. Group A now can receive $109 in rent, but at a cost of $10; therefore, their net value is $99. Their neighborly actions have left them worse off than before. Free riders, group B, are better off by $2, which they also receive from the $2 neighborhood benefit added to their $100 base, at no cost to them. Conversely, the payoff/loss when group B chooses to renovate, while group A does not, is shown in the lower left pair of cells of table 2.5.

Say that groups A and B both renovate and maintain their property. The payoff/loss pair of cells for this scenario is in the upper left of table 2.5. Both have a base rent of $100. Both expend $10 on their property. Both receive an increase of $2 because players in their group renovated, and both receive an additional increase of $2 from spillovers from players in the other group renovations. The rent for each group is now $111, but at a cost of $10, netting them as better off by $1.

Both groups are better off in the upper left pair of payoff/loss cells when all maintain and renovate. However, both groups have an incentive to break from this mutual action and become best off by not renovating, while the other players invest in renovations. All individuals in either group have an incentive to be best off; but, in their attempt to be best off; both groups of players are left in the equilibrium solution of the lower right pair of payoff/loss cells. The equilibrium solution is for neighborhood deterioration and disinvestments.

As seen elsewhere in this chapter, the driving mechanism for the payoff matrix of table 2.5 is the interdependencies and externalities that are a natural and always present component of real estate. You cannot control the actions of another, and yet others' actions affect your payoff. Consequently, the decisions you make must be made in the context of your expectations for the likely behavior of others.

There are a variety of situations where the normal geographic market forces of table 2.5 are circumvented:

- Everyone can collude to settle in the upper left payoff cell, where all are better off, but none is best off. Some neighborhood organizations are effective at obtaining that solution, while others are not. Some might argue that when such interdependencies result in such market failure and consequently in urban decline, that local government has an obligation to take corrective action (see table 2.4). Collusion can come about by restrictive neighborhood associations that require a minimum level of repair to maintain outward appearance. But the result of the collusion is that everyone is better off, yet everyone has an incentive to break with the collusion and become best off.
- Some older neighborhoods that have historical architectural value, combined with a desirable location, do begin the process of renovation. Often, such renovation occurs with the assistance of local government redevelopment initiatives, but generally only after significant decline has set in.
- Overcoming the incentive to be a free rider and view other's improvements as a cash cow to one's own net worth can be a costly and lengthy undertaking. In recognition of this, many governmental jurisdictions in Canada, Europe, and the United States, issue building permits only when the local planning council (see table 2.4) is satisfied that a consequence of development will not be the creation of an oversupply of real estate inventory, thereby setting off a downward spiral in the older real estate submarkets.

As the neighborhood declines in maintenance and value, successively lower income groups replace former occupants in what is referred to as the *filtering downward cycle*. Retailers depending on the consumption patterns of former residents either adjust to the changing geodemographics or go out of business. Commercial property values are thereby also affected and generally mirror changing geodemographics.

Gentrification is a process where higher income households rehabilitate a neighborhood, with the faith that others will join with them to achieve the upper left pair of payoff/loss cells (table 2.5). Gentrification displaces lower income households who earlier occupied the neighborhood and who themselves had formerly displaced higher income households residing in that submarket.

Succession of Property Types

Cycles can be measured in pecuniary terms, as was done above. But real estate deals with both land value and land use. As there are cycles of land value, there are also cycles of land use. What comes first, the chicken or the egg? What comes first, the grocery store or the residential dwelling? The real estate question is much easier than the farmer's question. The house comes before retailing. But what comes first in the succession of all property types and land uses?

In the industrial manufacturing era of the nineteenth century, the factory came first. The factory provided jobs. Labor would agglomerate near the factory. In that

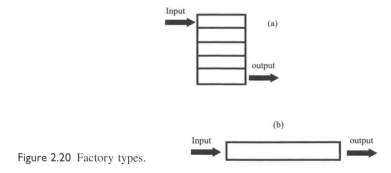

Figure 2.20 Factory types.

transportation-by-foot era, spatial equilibrium forces were as prevalent as now. Households would compete with one another for proximity to the place of work. This competition drove up population densities near the factories. These market forces then made residential land values highest near the factory node, and lower with distance from the node. After the threshold population was reached to sustain various types of retail, those with the lowest population requirements and selling the more frequently purchased items would enter the market first, to be succeeded by retail establishments of higher order as the population continued to grow with the job market. The factory then became the downtown. Retail establishments would be placed to intervene between the place of work and the place of residence, intervening between origin and destination.

Many urban factories were tall in the industrial age before electricity, as in the multistory representation of a factory building in figure 2.20. There was a high capital-to-land ratio. This architecture was adopted to conserve energy. Energy would be transported by way of belts on a pulley from a steam engine or water wheel to the machinery. The longer the belt, the more energy lost. So there was an incentive to agglomerate around the source of power.

Some nineteenth-century factory buildings like that in figure 2.20 still exist in the inner cores of American cities. Winston Salem, North Carolina, has some excellent examples. These factories have been converted into parking garages, retail shopping, condominiums, and warehouses. Nineteenth-century factory buildings in many of the larger U.S. cities are being converted to upscale residential lofts and offices.

Manufacturing, then, does not locate downtown. Nineteenth-century manufacturing facilities became the downtown of the then-new industrial-age cities. As the transportation-by-foot era gave way to transportation by automobile, well-paid workers took advantage of their greater mobility and moved their families away from the negative externalities of noise, pollution, and other possible environmental hazards of the industrial factory. This increased the geographic complexity of the city.

Manufacturing technology has changed as well. The modern factory building is more like the lower one in figure 2.20. There is an insignificant loss of electrical power over short distances, so there is no need to cluster production facilities around a central power source. The new factory has a low capital-to-land ratio.

Environmental regulations have also changed people's willingness to reside near industrial facilities. In the nineteenth century, workers resided nearby because they had to walk to work. Mid-twentieth-century workers used the auto to flee to the more environmentally friendly suburbs. However, spatial equilibrium forces may well compel the twenty-first-century worker to return near the place of work: national environmental regulations have mollified the negative effects of living near much of modern American industry. Instead, congestion of highways results in a great loss of otherwise productive time. Those centripetal forces pushing households near their places of work might be greater in the twenty-first-century than the centrifugal forces propelling them away.

Because of the change in technology, transportation, and economic base of many cities, the twenty-first-century city is very different from its nineteenth-century counterpart. The succession of land use today generally begins with either a node of job creation or some natural or man-made amenity, like Colorado's renowned Vail Valley. Jobs are concentrated at some location. Once sufficient housing and jobs are in place, retail and services follow. Retail and service industries are the fastest growing sectors of the United States today. Those employees are housed, providing the threshold opportunities for more retail and services. This multiplier effect was discussed earlier in this chapter.

But what about those who move to enjoy the amenities of the high mountain landscape of Colorado, or the lovely winter weather of Florida? The engine that drives amenity markets are either households who bring their wealth that has been created elsewhere with them, or "commute to work" via modem. This introduces the next section on information technology.

Information and Technology

Did President Eisenhower understand what effect the National Highway Defense Act would have on the American city when he signed the act that funded the creation of the nation's Interstate Highway System? Apart from the limited-access German Autobahn, Route 66, and the Lincoln Highway, there was no precedent. The construction of the Interstate Highway System was the largest public works program in the history of the planet, the two closest rivals being the Great Wall of China and the ancient Egyptian pyramids.

Most have come to believe that the early twenty-first century is the dawn of the information age. Do we know how information technology will change the city? How will information technology change the urban built environment, and real estate itself? There are guesses, and there is some evidence (Thrall, 1998, Miller 2000). But, as with the interstate system, we may need to wait and see how the information technology actually changes the city.

An increasing portion of the U.S. labor force is engaged in work that allows them to live most anywhere so long as they stay in electronic communication. They can choose to live in a Florida-like environment in the winter, and a Vail, Colorado-like environment in the summer. They can choose whatever combination of natural and urban-built amenities they prefer. Their numbers are expected to increase.

Some workers may have the desire to be free of a centralized workplace, and some of those are able to maintain or even increase their levels of productivity when not so tied. Employers have also found that they can be better off if some employees can productively occupy office space elsewhere that they do not have to pay for. The distributed work environment, where the workforce is connected electronically, has the potential to dramatically change office markets. Some businesses have adopted "hoteling" of office space, where an employee reserves a desk for an hour or a day when they must be physically at the workplace. Hoteling is not yet pervasive among office market renters, but through time it may significantly reduce the required allotment of office space per worker.

The Internet allows a workforce to be dispersed away from centralized office environments. Some have argued that the Internet has the potential to make office and retail markets obsolete, just as those nineteenth century factory buildings discussed earlier became functionally obsolete with changing technology (Baen 2000, Baen and Guttery 1997). However, others do not believe that the Internet will anytime soon lead to the demise of retail and office centers (Roulac 1994). Although the Internet has allowed some to have greater choice in where they live and shop, not all with the same locational freedom are making the same choices. At the time that the Internet is allowing greater dispersal of populations, the hottest real estate submarket in the United States and Canada is the historic downtown central business core. What makes the downtown core so desirable today?

The Internet has not displaced all need for face-to-face interaction. But the Internet has increased the productivity of labor, so much so that those whose productivity has been so greatly enhanced by the Internet view commuting for face-to-face meetings as too costly and too great a loss of productivity. They might have the strongest preference for locating at central downtown locations.

The structure of the family unit has changed. Today more than 55 percent of married persons become divorced within six years. Although the desire to be happily married has not declined, the reality of the low probability of a long lasting marriage has lowered the rate of marriage for all marriageable cohort groups. Changing family structures have changed the requirements for housing. Among unattached persons, there is a desire for locating in a place where there is a high likelihood for encounters that lead to successful, albeit more transitory, relationships. Downtown is the place to go to meet. The suburbs do not offer the same opportunities. Many suburbanites do not even know their neighbors, in part because the architecture of the suburbs discourages interaction. To remain competitive, the "lifestyle mall" (see chapter 7) is intended to accommodate this exchange in the suburbs. But there remains the reality that the newly divorced, especially with children, have a substantial reduction in finances. The locational patterns and choices of lower income people have been discussed earlier in this chapter; the same analysis would hold for many newly divorced.

In the year 2000, Coldwell Banker announced that 30 percent of its homebuyers first viewed the houses they purchased on the Internet, and many of those even engaged in cyberspace tours of those houses. The Internet has become an agent of change in the real estate industry. The Internet will be affecting real estate in more ways than directly affecting home sales. The Internet allows people who are greatly dispersed to provide a threshold demand for specialized provision of goods and services, which heretofore might only have been available in the largest cities of the world. The Internet makes those goods and services available to consumers residing in the smallest towns. These effects of the Internet combine to make living outside of large urban agglomerations possible, while at the same time enjoying the benefits of big-city shopping. It is not known yet whether the Internet can be a substitute for "shopping as entertainment (Underhill 1999)."

The Internet is a new twist on an old idea: mail order. Many pundits claimed the end of the modern retail store when Sears published its first mail-order catalog in the nineteenth century; more than 100 years later, people still shop at stores. But the Internet is truly different from a large mail-order catalog. First, there are many vendors, with very low cost of entry. Second, the catalog can be continuously updated. Third, the images can be moving and can also include sound, making the catalog seem antiquated. Perhaps most important, the Internet allows communication between customers, not just between the seller and the buyer.

Customer-to-customer communications have become commonplace on Internet bulletin boards. There consumers of, say, high-end espresso machines can ask which machine they should buy, and from whom, and who has the best price and service. A displeased catalog mail-order customer in rural Kansas may choose not order from a firm again if displeased. Satisfied customers might place a second order. Via the Internet, that same displeased or satisfied customer can inform the global population of the quality of service and their level of satisfaction. The Internet promises to bring perfect information to the consuming market. And those consumers are not geographically locked into making their purchases based on geographic principles of distance minimization. Instead, the decision to purchase can be based on delivered price, product, and service. "Big box retail" can be considerably affected during the first decade of the twenty-first century (Baen 2000, Baen and Guttery, 1997).

The Internet allows opportunity for dispersal, just as the interstate highway allowed opportunities for dispersal. But people were different in the 1950s than they are in the 2000s. The Internet will not necessarily lead to a decline in the city, as other forces are in place to keep cities viable. Spatial equilibrium centripetal forces for locating at urban core places might be greater than the centrifugal forces leaving to population dispersal. It is too early to tell which force is stronger.

Concluding Remarks

Cities, beginning with the first trolley cars, automobiles, and now the Internet, have become increasingly complex. We do understand the trajectory that improved modes of transportation set forth for the city, but we still do not know what changes will be brought about by the Internet. If cities were as simple as they were in the nineteenth century, there would be no need for the business geographer's market analysis, and market analysts would be out of a job.

An understanding of urban form and urban morphology requires a general descriptive study of the forces that shape a city. This chapter described those general forces. Market forces were grouped into nine categories: geodemographic, transportation, interdependencies, spatial equilibrium, population change and relocation, government, economic base, real estate cycles, and information. Chapter 3 goes further and explains how changes in these general forces translate into changing land use and land values.

3

Unifying Urban Land Use and Land Value Theories

The market analysis report that is submitted to the decision maker (see chapter 4) should include a descriptive, qualitative overview of the context the real estate project has to the existing and changing urban environment. To better accomplish this task, one should have knowledge of the general theory regarding the market processes that bring about land use and urban form. This chapter presents three relevant general theories of land use and land value. Together, these general theories provide a general qualitative understanding of how the existing urban environment came to be and allows the analyst to prognosticate the trajectory of change of urban land uses and land values.

The first two of the general theories presented here arose out of an attempt to explain agricultural land values and land uses. Why should a discussion of the agronomy sector be included in a book on urban real estate analysis? First, all the general theories relevant to land values and land use, and their spatial distribution within a city, are part of an intellectual heritage that dates from general theories of agricultural land values and land use. Second, much of new urban development occurs at those suburban margins. To understand development at the suburban margins, there must be an understanding of the nonurban land uses and land values at those locations. The third general theory explains spatial equilibrium and its role in shaping urban land values and land uses.

Two eighteenth-century theorists, David Ricardo and Johann Heinrich von Thünen, are credited for having created a vast and sometimes opposing literature on land valuation. Ricardo's economic theory was based upon the relative productivity of sites.

In contrast, von Thünen's geographic theory was focused on the locational component of land values and land use. The juxtaposition of these two competing giants of land theory in many respects still differentiates economists and geographers even today. After the theories of Ricardo and von Thünen are presented, an overview of the consumption theory of land rent (CTLR) is provided. The CTLR is my general theory and methodology for evaluating urban housing land use, land values, and urban form (Thrall 1980, 1987, and see 1991).

David Ricardo

Ricardo hypothesized that high rents were attributable to the "niggardliness of nature"—in other words, to scarcity. Furthermore, rents were related to the "original and indestructible power of the soil"—in other words, to differences in the productivity of land. He believed that the most fertile lands are put to use first, with production extending to less favorable lands only as demand justifies this (quoted in Berry et al. 1987, p. 222).

Ricardo's primary argument was that the most productive land would have the highest land value, or land rent. The next most productive land would have land rent equal to the most productive land, less the value of investment that would be required to bring that land up to the most productive level. In other words, the differential in the value of productivity between the two land parcels is subtracted from the value of the more productive land to arrive at the value of the less productive land. In a Ricardian world, land value then depends on absolute and relative productivity of land. Table 3.1 gives a numerical illustration.

In table 3.1, the base rent for the most productive class A land is $100. That value is then used as a comparable for class B land that is the next most productive land. However, $10 is required to bring class B up to be equal to class A land, so the rent or value of class B land is $100 − $10 = $90.

Ricardian land rent theory remains central to estimation of the value of nearby land based on comparable sales and cost differentials of what is required to bring the lesser property up to being equal to the superior property. Ricardian land rent theory is at

Table 3.1. Numerical illustration of Ricardian rent

	Class A land	Class B land
Base rent	$100	$100
Required additional investment to bring property up to highest level of productivity	$0	$10
Property value	$100	$90

the heart of contemporary real estate appraisal. A shortcoming of Ricardian land rent theory is that it does not account for location.

Johann Heinrich von Thünen

Whereas Ricardo focused on absolute and relative productivity of land, von Thünen's focus was on absolute location and relative spatial location. Von Thünen hypothesized that the value of land arises from the bidding process of the market. There are those who are owners of land and those who are bidding to use that land. The goal of the landowner is to maximize returns. Land users bid against one another for the right to use land. Von Thünen believed that market price for the use of land could be expressed in mathematical form as:

$$\text{Land rent} = (\text{total revenue}) - (\text{total cost}) - (\text{total transportation cost}). \quad (3.1)$$

Total revenue is how much the land user receives from the sale of their product at the central market. Total cost includes the necessary agricultural inputs as well as *opportunity cost* of the land user. Von Thünen is credited with being the first to use the concept of opportunity cost, which is the alternative highest returns that may be received from some other available activity.[1] Transportation cost is the total expenditure of getting the goods to market.

Total revenue to the land user is the same for every land user who produces the same amount of crop. In other words, the price received at the market per unit of agricultural output will be the same regardless of where production takes place. As with Ricardo, total cost can vary with the differences in the productivity of land; but to keep the discussion short, assume as von Thünen did that land everywhere in the relevant market has the same level of productivity. The resulting assumption of an undifferentiated plain is referred to as the *isotropic surface assumption.*

Transportation costs are calculated as a multiple of:

$$(\text{quantity shipped}) \times (\text{shipping rate per quantity per distance}) \times \\ (\text{distance shipped}) \quad (3.2)$$

If 100 pounds of apples are produced and shipped, and if the shipping cost is \$0.20 per pound per mile, and the orchard is 10 miles from the market, then the total transportation cost is $100 \times 0.20 \times 10 = \$200$. At the market where distance is zero, the transportation cost is also zero. At the market center, the total land rents (land values) are merely the difference between total revenues and total costs.

The general message from von Thünen is that loss of accessibility, as measured by rising transportation cost, decreases land values. As distance increases from the desired place of access (i.e., the market center), the amount paid in transportation costs also increases. As transportation cost increase, and since revenues received from the sale of the agricultural product at the market are constant and therefore not dependent on the origin of the product, then land rents (land values) decline as access to the market diminishes.

Von Thünen introduced the concept of *highest and best use*, meaning that the market bidding process will bring about the most productive use of land becoming the prevailing land use, and that those land users outbid other uses of land at that location. In other words, the land user that is willing and able to bid the greatest

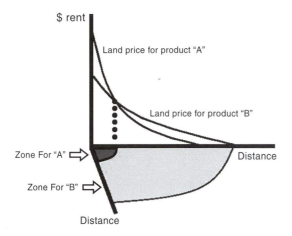

Figure 3.1 Von Thünen's concentric ring theory.

amount will receive the authorization from the landowner to produce on the land. Geographically, the market bidding process among competing uses of land gives rise to the concentric rings of land use, as shown in figure 3.1.

A two-product world has been assumed for figure 3.1. Land values decline with increasing distance from the market for both agricultural products. Those who choose to produce product A are willing and able to pay a greater land rent near to the market center than producers of product B. Going away from the market center, that situation prevails until the dotted line where both producers of A and B are willing and able to pay the same rent. For distances beyond that location, producers of product B outbid producers of product A for the entitlement to use land. Because land owners contract with the highest bidder, then two concentric zones of land use are formed around the market center: zone for product A and zone for product B. This is known as *von Thünen's theory of concentric zones*. Von Thünen's general theory can be extended to any number of competing land uses.

A limiting characteristic of von Thünen's land rent and land value theory is that it was developed to describe, explain, and predict land values and land use in the agricultural sector: land users and land owners making decisions as to what to grow where in order to maximize profit. Instead, our goal is to describe, explain, and predict urban land values, particularly those land values that come about because of households' decisions to consume.

Grant Ian Thrall

An introduction to a general theory of the value of urban land for housing is presented here. The general theory integrates and extends the concept of spatial equilibrium that was introduced in chapter 2 and the economists' microeconomic consumer behavior theory (Thrall 1980, 1987). The general theory can accommodate what-if scenarios as there are changes in the values of the various determinants of land value and land use, some of which were described in chapter 2, including changes in (Thrall 1987):

- Transportation costs
- Spatial distribution of urban activity centers (i.e., multinodal city)
- Income levels and income distribution
- Taxation
- Planning, zoning, and rent controls
- Public goods and externalities.

Thrall proposed two general theories of land values and land rent for the urban environment to deal with the two predominant uses of land in the city. His *consumption theory of land rent* (CTLR), summarized below, deals with land values as they arise in the housing sector. His *production theory of land rent* (PTLR; Thrall 1991), deals with land values as they arise in the commercial sector, including retail, services, and industrial,[2] and is not presented here.

As is the case in microeconomics theory of consumer behavior, Thrall's CTLR proceeds in sequence with (a) derivation of a budget constraint and (b) specification of a set of utility curves. Because of the importance of location to real estate market analysis, the microeconomic consumer behavior theory must be expressed to also conform to the geographer's notion of spatial equilibrium. Therefore, the next step is (c) specification of a spatial equilibrium. When put together, the preceding steps then allow for (d) derivation of spatial equilibrium land values across space, (e) derivation of spatial equilibrium consumption of land as well as a general set of consumer goods across space, and (f) identification of the new urban growth boundary, thereby establishing a general theory of a city's primary trade area.

Because the CTLR has its foundations in both the geographer's spatial equilibrium and the same microeconomic theory that provides the foundations for the law of downward sloping demand, and because the modifications to microeconomic theory are reasonable so that locational access and spatial equilibrium can be included in the analysis, the results of the general geographic theory are equally valid as the results of the general microeconomic consumer behavior theory.

The CTLR provides a context within which all these concepts are integrated, related, and intrinsically bound together, which is a necessary component to the analyst's understanding of the general urban market.

The Budget Constraint

A household's disposable income at distance s from the important place of access, such as the CBD (central business district), is that amount of income remaining after transportation expenditures have been accounted for. Disposable income is used here to mean the difference between transportation costs, k, per unit distance s from the CBD, and gross household income y; namely, $y - ks$. A linear transportation cost function is assumed. Total transportation costs are ks.

We are interested in how households allocate their income between land expenditures and all other expenditures. Therefore, the collection of all other expenditures will be agglomerated and are referred to as the *composite good*. Let p be the price of composite good, z be quantity of composite good, r be price of land, and q be quantity of land. Because expenditures on land and composite good must equal disposable income, we can write:

$$y - ks = pz + rq. \tag{3.3}$$

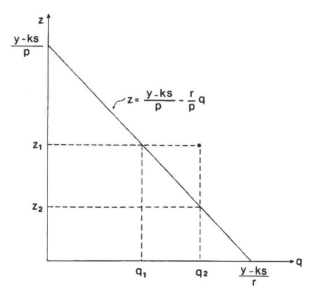

Figure 3.2 Budget constraint (after Thrall 1980, 1987).

Equation 3.3 is the household's budget constraint as commonly seen in microeconomic consumer behavior theory, except for the inclusion of transportation cost.

The budget constraint is graphed in figure 3.2, using the two-point method for a straight line. The intercept on the ordinate or composite good axis is:

$$z = (y - ks)/p \qquad \text{(given } q = 0\text{)}, \tag{3.4}$$

and on the abscissa or quantity of land axis is:

$$q = (y - ks)/r \qquad \text{(given } z = 0\text{)}. \tag{3.5}$$

Equations 3.4 and 3.5 are the quantity of z and q respectively that the household can afford to consume if the entire disposable income is devoted to one of those two products. Also, equation 3.3 can be rearranged to be in the form of a straight line:

$$z = (y - ks)/p - (r/p)q. \tag{3.6}$$

The budget line in equation 3.6 has intercepts shown in equations 3.4 and 3.5, and a slope $(-r/p)$ equal to the negative ratio of land price to composite good price. Equation 3.6 is graphed in figure 3.2.

A household can afford to consume any combination of z and q on the budget line, and within the area bounded by the budget line. We are restricting the household to spend all their income, and we are constraining them not to borrow. The additional complexity of savings and borrowing can be added to the analysis, as the CTLR is robust; but for our purposes in this introduction to the general theory, doing so would just add a level of complexity without any real benefits to the analyst.

The Utility Frontier

The utility (or indifference) curve is a set of points, each of which represents a combination of goods that the household is indifferent to. However, households are not

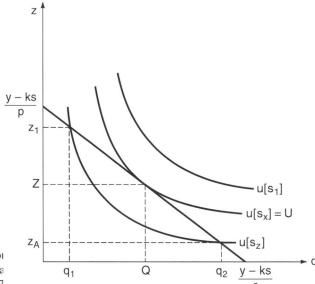

Figure 3.3 Equilibrium consumption of goods (after Thrall 1987

indifferent to combinations of goods that are on higher, or lower, curves. Higher curves are preferred to lower because they offer an opportunity to consume more of at least one of the goods. Figure 3.3 shows the utility and budget curves together. In figure 3.3, the household can afford to consume more z than the equilibrium amount, Z. However, by doing so, the household must balance z with decreased expenditures on q. The combination of Z and Q shown in figure 3.3 is the economics equilibrium demand for composite good and land, when land price is r, composite good price is p, income is y, and transportation cost is ks. However, land price is unknown, and there are many distances, s, from the important node of access, such as the CBD.

Spatial Equilibrium and Land Values

At a Specified Location

Land value, or land rent, is an unknown price. Let U from figure 2.13 be the spatial equilibrium level of household welfare. In figure 2.13, U is mapped over space. In figure 3.4, that same level of welfare is graphed in terms of z and q.

In figure 3.4, if land prices were higher than $R[s_1]$, say r', then households would only be able to afford level of welfare u', which is lower than the spatial equilibrium level of welfare required by the market. If land prices were lower than $R[s_1]$, say r'', then households would be on a level of welfare above the spatial equilibrium, U. Land price at s_1, $R[s_1]$, satisfies spatial equilibrium conditions. With all the prices known at distance s_1 from the CBD, the spatial equilibrium consumption of land, $Q[s_1]$, and composite good, $Z[s_1]$, become known as shown in figure 3.4.

Consider the geographic consequences of welfare (utility curve) being U', which is lower than the spatial equilibrium, U (see figure 3.5). For there to be both economic

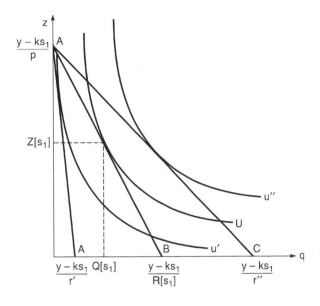

Figure 3.4 Differences in two welfare schedules (after Thrall 1980, 1987).

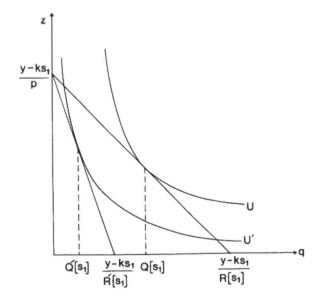

Figure 3.5 Comparison of two spatial equilibrium welfare levels (after Thrall 1980, 1987).

and geographic equilibrium, land prices would be higher at $R'[s_1]$, and land consumption would be less at $Q'[s_1]$.

Across All Distances

To demonstrate, derive, and prove the behavior of land values over space, first consider the possible alternatives. Land prices could be the same everywhere. Land prices could increase with increasing distance from the CBD. Land prices could decrease with increasing distance. We will consider each of these possibilities in turn.

Can land prices remain the same as land prices are at s_1, yet with increasing distance from the node? If distance increases, then transportation expenditures, ks, must increase, and disposable income will decline. Holding land values constant for all locations and equal to $R[s_1]$, then the two intercepts shown in equations 3.4 and 3.5 must also decline. Since we are investigating if land prices can be the same everywhere, then the slope, $R[s_1]/p$, does not change. Hence, the budget line in figure 3.6 has a parallel drop from AA to BB. That places the household on a lower level of welfare, like that discussed in figure 3.5. So spatial equilibrium conditions are violated. In other words, we have proven that just as there is a law of downward sloping demand, land prices cannot be the same everywhere, all other things being equal.

Can land prices increase with increasing distance from the node? Transportation expenditures increase with increasing distance from the node. Disposable income declines moving the intercept on the composite good axis down to point B in figure 3.6. If land prices had remained constant, the intercept on the land axis would be point B there as well. Because land prices are proposed to increase, then the intercept on the

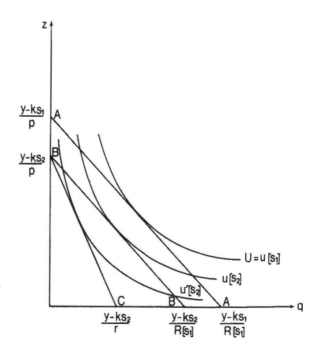

Figure 3.6 Land rent is neither constant nor increasing with increasing distance from the city center (after Thrall 1980, 1987).

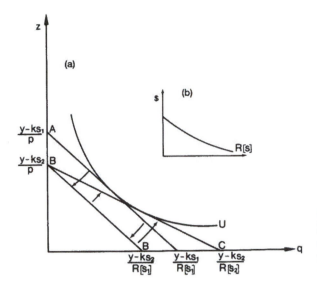

Figure 3.7 Spatial equilibrium land values decrease with decline in accessibility, all other things being equal (after Thrall 1980, 1987).

land axis must be left of B, say C. That places the household on an even lower level of welfare than was the case with land prices being the same everywhere. Again, spatial equilibrium conditions are violated. In other words, just as there is a law of downward sloping demand, we have proven that land prices cannot increase with increasing distance from the CBD, all other things being equal.

The only possible alternative that remains is that land prices decline with distance from the CBD. In figure 3.7, land prices are held constant as distance increases from s_1 to s_2. That gives budget line BB. Allowing land prices then to decline until the budget line is tangent to the spatial equilibrium level of welfare, U gives rise to the spatial equilibrium level of land prices at s_2, namely $R[s_2]$. Therefore, we have proven that land prices must decline as shown in figure 3.7b.

The procedures followed for figure 3.7 can be repeated for any number of distances from the node. In figure 3.8, spatial equilibrium land prices are derived for three locations, $s_1 < s_2 < s_3$. The spatial equilibrium consumption of land and composite good can then be derived as the tangency between the three budget lines and the spatial equilibrium level of welfare, U.

As seen in figure 3.8, $Z[s_1] > Z[s_2] > Z[s_3]$, so the spatial equilibrium consumption of composite good declines with increasing distance from the node. This can be referred to as the *space demand curve* for composite goods. Also, from figure 3.8, $Q[s_1] < Q[s_2] < Q[s_3]$, proving that spatial equilibrium land consumption increases with increasing distance from the node. Because land consumption and population density are the inverse of one another, then we can say that population density declines with increasing distance from the CBD, or other important node. This empirical result was described in chapter 2. Here, using Thrall's general CTLR, that same empirical result of chapter 2 has been proven. Again, just as there is a law of downward sloping demand, we have proven that land prices and population density must decrease with increasing distance from the CBD, all other things being equal.

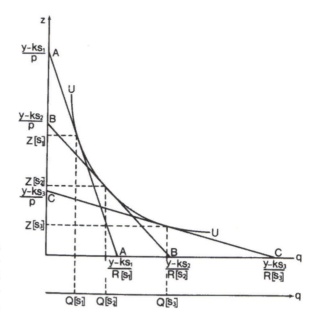

Figure 3.8 Spatial equilibrium land rent and spatial equilibrium consumption of land and composite good over three locations (after Thrall 1980, 1987).

Since at location s, quantity of land, $Q[s]$, and population density, $D[s]$, are the inverse of one another, we can write:

$$D[s] = n/Q[s], \tag{3.7}$$

where n represents the number of people in a household, so density is measured in units of household density. Therefore, as land consumption increases, population density decreases. So population density in accordance with chapter 2 declines with increasing distance from the city center, all other things being equal.

Space, the Missing Dimension

The spatial equilibrium solution for three locations was demonstrated in figure 3.8. That solution will now be extended for a continuous space, enveloping $s_1 < s_2 < s_3$.

In figures 3.2 through 3.8, the geographic dimension has been collapsed to only represent some sample locations. Instead, the geography dimension of these figures extends toward and away, making those figures three-dimensional. This is shown in figure 3.9. Figure 3.9 is the same as figure 3.8, but includes geographic space. The spatial equilibrium level of welfare is a plane in figure 3.9. The spatial equilibrium space demand curve for land and composite good can be seen here in figure 3.9 to be smooth and continuous. Next, the spatial equilibrium geographic boundary to the urban built environment will be derived.

Urban Limits

Where is the boundary of the city? This was explored in chapter 2 (see table 2.1). Thrall (1980, 1987) provides several scenarios linking this discussion to table 2.1 and

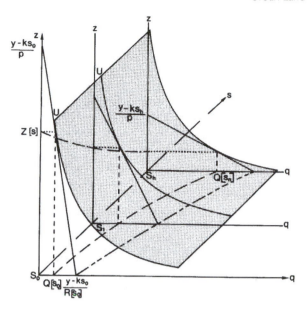

Figure 3.9 Adding the spatial dimension and spatial equilibrium (after Thrall 1980, 1987).

to the general market processes. Here the derivation of urban limits will be limited to only what Thrall in his CTLR defined as an *interactive city*. An interactive city is one whose boundary is defined by interaction between the urban and the nonurban agricultural market, as shown in figure 3.10.

Land values have been proven to decline with distance from the CBD. The question arises, is the urban boundary defined where some other market is willing and able to pay a higher land value or rent, as in Thünen's two-product world, except where one product is urban land use and the other agricultural? Thrall (1979) demonstrated that agricultural land values, $\rho[s]$ do not have the largest role in establishing the geographic boundaries of a city. Instead, Thrall reasoned, a city is constrained laterally by the cost of converting nonurban land to urban land, $\delta[s]$. Refer back to the discussion related to table 2.4, and the role of the government in providing urban infrastructure. The variable $\delta[s]$ is the cost of urban infrastructure. The slope of $\delta[s]$ may be an increasing, constant, or decreasing function of distance, depending on the cost of installing infrastructure near to the urban boundary or far from it.

Among the questions that were raised in chapter 2 are, who is to pay, what infrastructure is to be created, and where is that infrastructure to be created? These questions arise again in figure 3.10. If infrastructure is viewed as a free good (i.e., the government or somebody else pays), and if that infrastructure is installed, then the radial boundary of the city will be at h^*. That does not necessarily mean more people will inhabit the city; rather, the same number of people can be dispersed over a larger area.

Among the consequences of dispersing persons over a larger area are the decline in the retail sector of the city, since it has been proven earlier in this chapter that composite good consumption decreases with increasing distance from the city center, all other things, including income, being equal. One explanation for this is that the

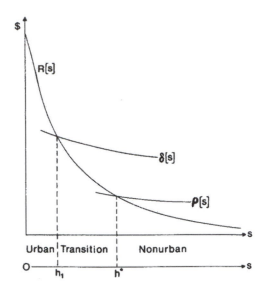

Figure 3.10 Urban limits as defined by the cost of converting nonurban land to urban land (after Thrall 1980, 1987; also see Thrall 1979).

more dispersed households are, the greater the allocation of a household budget to transportation expenditures, and time involved in accessing distant places of shopping. This is discussed more fully in chapter 4. Land investors will be most active in the zone labeled *transition* because, if government can be convinced to intercede and pay for the cost of conversion, then land there becomes developable. Urban households are willing and able to pay only $R[s]$, and beyond h_1 urban households are not willing or able to pay for the conversion price, $\delta[s]$. Much of the politics of real estate development deals with $\delta[s]$ in the transition zone. If $\delta[s]$ is somehow paid for by others, or by way of zoning the infrastructure is determined not to be necessary, then land prices will rise from agricultural land, $\rho[s]$, to instead $R[s]$.

Bringing It Together

Population density was described in chapter 2 as declining with distance from a node, such as a city center. In this chapter, population density has been proven to decline with distance from a node. Land values have also been proven in this chapter to decline with increasing distance from the node. And the cost of conversion has been demonstrated as being capable of geographically binding the city radially. These concepts are brought together in figure 3.11.

Population density in figure 3.11 is drawn as three dimensional, extending all the directions of the compass. The volume of the area under the population density cone is the total population of the city (for discussion of open and closed cities, see Thrall 1987. Geographic information systems (GIS) technology makes the above discussion, particularly for that of figures 3.9 through 3.11, exceedingly relevant for the real estate market analyst. Before the recent technological advances of GIS, specifically the Census Bureau's publication of U.S. streets in digital format (known as TIGER/Line files) in 1990, real estate market analysts had only glimpses of what the general theory like

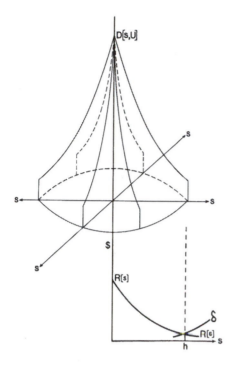

Figure 3.11 Spatial equilibrium population cone.

that presented in this chapter proposed. Now, with GIS, we can visualize population density, expenditures on composite good, and land and housing value surfaces. The general theory explains what we see in such visualizations, and helps us anticipate trends. GIS-based spatial statistical procedures in combination with the general theory allow us to fill in the gaps where we do not have spatial data and to project what these surfaces will likely be in the future.

Exogenous Changes

What if there were a change, or there is an anticipated change, in transportation costs, location or number of urban activity centers, income levels and income distribution, taxation, planning, zoning, rent controls, public goods and externalities, and so on? What would be the effect on land values, population density, total population, composite good consumption, urban radius, and so on, as described above? The answer provides an expectation as to the trajectory of how the market will change, and therefore the trajectory of land values, density, and so on (see figure 3.12).

Begin with the system in spatial equilibrium. Next we ask if the exogenous change in the phenomena is to be interpreted as a budget constraint shifter or a utility shifter. The phenomena whose affect is being tracked directly affects either the budget constraint or the households' utility curves. Examples of budget constraint shifters include changes in income, transportation cost, and taxes. Examples of utility shifters include changes in time required for commuting, zoning, externalities, public goods, and so on.

(a)

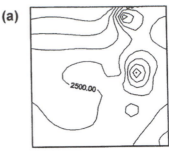

(b)

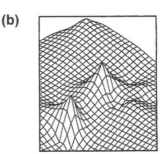

Figure 3.12 Hypothetical three-dimensional contour.

We first calculate the *direct effect* (Thrall 1987), namely, how the household's welfare surface changes because of the external effect, while holding spatial equilibrium land values constant. For example, if transportation cost declines, then households at the city center (CBD) will not be significantly affected because they do not commute. However, the budget of households near the urban periphery will increase substantially because a large proportion of their budget is allocated to transportation expenditures. The new higher budget allows households to afford higher welfare. Therefore, the direct effect on household welfare across space because of a decrease in transportation cost will be as in figure 3.13. Utility now available to households rises with distance from the CBD. The system has been propelled out of spatial equilibrium.

We next calculate *indirect effects* (Thrall 1987). Indirect effects are what results as land values change as spatial equilibrium is restored. The time required to restore spatial equilibrium might be short or might be several decades. How land prices adjust as the system returns to spatial equilibrium depends on the rate of in- and out-migration. If a city is susceptible to high levels of population change, then the city is referred to as an open city. If the population of the city does not change, then the city is referred to as a closed city. Indirect effects will differ between open and closed cities.

An *open city* is defined as one where the population can change and spatial equilibrium welfare is constant. A *closed city* is defined as one where the spatial equilibrium welfare can attain a new higher or lower level and where population is constant. Both polar solutions are normally calculated, thereby establishing an envelope around the possible solutions.

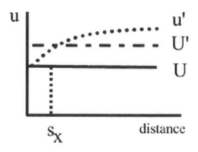

Figure 3.13 Direct effects on utility surface from decrease in transportation cost (Thrall 1987).

For an open city, land prices will adjust to bring welfare at all locations back to the initial spatial equilibrium level, namely U. Because direct effects do not affect households at the CBD, then land prices at the CBD will not change (figure 3.13). Because land prices do not change, then the demand for land will not change either at the city center. However, toward the urban periphery, land prices will rise to counter the increase in welfare there. As households vote with their feet, land prices are driven up, as is population density. Welfare then moves back to U. See figure 3.14.

For an closed city, land prices will adjust to bring welfare at all locations to a new spatial equilibrium level, say U', where $U' > U$. In spatial equilibrium, welfare must everywhere change to U'. But direct effects did not affect households at the CBD. Welfare of households near the CBD did rise, but not by an amount equal to the new spatial equilibrium, U'. See figure 3.13. Land prices then fall as compensation to households interior to s_x. At s_x, land prices remain the same because the direct effects elevated equilibrium welfare there to the new spatial equilibrium level. However, the direct effect elevated welfare at locations beyond s_x to be above the new spatial equilibrium U'; therefore, the trajectory for land prices at those locations is to increase and thereby negatively compensate those households with welfare being driven back down to the new spatial equilibrium. As households vote with their feet, land prices and population density declines at the CBD. At the urban periphery, land prices and population density increase.

Figure 3.14 shows an envelope of possible changes around the initial spatial equilibrium levels, which were calculated above. The envelope is formed by the solutions for open and closed cities. Depending on the rate of in- and out-migration, land prices can either decline at the city center or remain the same. Land prices will rise at the urban periphery. Market forces support either a decrease in population density at the city center or no change at all and at the urban periphery support an increase in population density.

Direct and indirect effects can be calculated for a variety of changes in exogenous phenomena that affect the urban market. Changing exogenous phenomena include both budget constraint shifters and utility shifting phenomena (see Thrall 1987).

Figure 3.15 summarizes a sample of various direct effects on household welfare that result from a variety of common exogenous change. For the proofs, and for other changes, such as zoning, see Thrall (1987). The real estate market analyst must keep in mind that it is how the exogenous change directly affects welfare and how that

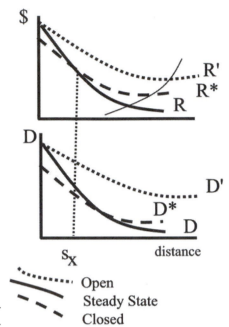

Figure 3.14 Envelope of direct and indirect effects (Thrall 1987).

welfare changes as the system is brought back into equilibrium that ultimately change land values and population density.

In figure 3.15, the system begins at spatial equilibrium with welfare surface U. The exogenous change propels the obtainable welfare to the curved dotted line, denoted as u' and u''. In an open city, people vote with their feet, moving in and out of the city, resulting in changes in land values and population density, and welfare equilibrating at the original level, U. In a closed city, the population may move about within the city until a new spatial equilibrium welfare level, U', is arrived at. If the dotted line is above the final spatial equilibrium welfare line, then land prices will increase there. If the dotted line is below the final spatial equilibrium welfare line, then land prices will decrease there. Where the lines intersect, there will be no change in land values.

Conclusion

In the development of our intellectual thought, particularly that for business geography and real estate market analysis, general theory until recently had a substantial edge over data analysis. The rise of technology, particularly GIS, has shifted that edge to data analysis. However, there is still a very substantial role for general theory. General theory is used to identify the problem. It binds elements of the problem together. It allows us to anticipate what-if scenarios, even if events of a particular type have not

Changing Exogenous Factor	Graph of change in direct welfare, in contrast to final spatial equilibrium welfare
• Increase in income • Decrease in income tax • Increase in mortgage interest deductibility • Decrease in transportation cost • Increase in public good land compliment	(a) 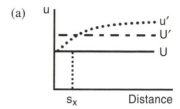
• Increase in public good land substitute	(b) 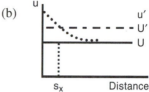
• Decrease in property tax rate	(c) 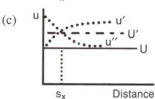
• Introduction of a superior transportation access node at S_v	(d) 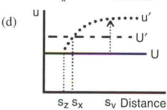

Figure 3.15 CTLR direct effects on household welfare resulting from a sample of exogenous changes (based on Thrall 1987).

occurred in the particular market. General theory adds the dimension of explanation to description.

In this chapter, three general land use and land value theories were presented: that of David Ricardo, Johann Heinrich von Thünen, and Grant Thrall. Depending on the circumstances, each can be drawn upon to explain and to anticipate how market forces will change because of some external influence.

4

Conducting Real Estate Market Analysis

This chapter establishes a general framework for conducting market analysis for all types of real estate. The subsequent chapters show how, with appropriate modifications, this general framework applies to specific real estate product types. The following chapters also include examples of the general approach introduced here.

Four general steps must be included in all real estate market analysis. This chapter will progress through each of the four steps. First, a trade area, also known as a market area, must be established from which to draw the data for the real estate market analysis. Explanation is provided on how to delineate a trade area for the project, including how large the trade area is and the geographic delineation of its boundaries. Second, to evaluate the competitive position of the project, competing supply is estimated. Competing supply includes both current and projected supply within the trade area derived in the first step. To identify competitive projects, the market analyst must also determine what segment the real estate project is to compete in. Third, demand must be measured. Demand estimation includes the assembly and use of projections of economic, geographic, and social indicators that together influence the demand for a specific real estate project. Occasionally, data readily available from commercial data vendors are inadequate for specific types of real estate projects. In these circumstances, demand is estimated using primary research, including surveys and focus groups, as well as assembly of primary data.

The fourth step is compiling the report and presenting the analysis to client. The report must reconcile the results of the foregoing three steps with the goals and needs of the client. The client may be an investor, a developer, a redevelopment agency, and so on. The objectives of the client might influence how the fundamental first three steps are interpreted. The market analysis report generally concludes with recommendations, as well as a description of the overall project within the wide context of the economic, geographic, and social forces that are shaping the urban built environment. The analyst draws conclusions, including projecting absorption rates and pricing recommendations. The organization of this chapter follows the steps listed above. The general background for each step is provided.

Step 1: Procedures for Establishing a Trade Area

The first step, establishing a trade area, includes four parts. (1) The definition of a trade area is closely tied to the Applebaum analog and customer spotting methods, which are the most commonly applied methods of trade analysis. The concept of trade area is defined, and these two methods are explained. (2) An overview of regression analysis is presented because a basic understanding of this statistical procedure is required to implement the analog method. (3) The analog procedure for revenue estimation and sales forecasting is demonstrated using regression analysis. (4) Three general approaches for delineating the geographic boundary of a trade area are presented. The three general approaches are rules-of-thumb, gravity and spatial interaction models, and customer surface modeling.

Definition and Uses of Trade Areas

A *trade area* is a geographic region from which a real estate project draws most of its customers (Ghosh and McLafferty 1987; Thrall and McMullin 2000b).[1] The population within the trade area has the greatest probability of engaging in exchange with the project being evaluated. Outside a trade area, the population is expected to have a lower level of interaction and exchange with the project. Within the trade area, customers of the real estate project are generally in greater geographic proximity to one another. The converse is expected for customers outside the trade area. Because analysis of a trade area is inseparable from analysis of phenomena and information on a map, application of geographic analysis and geographic technology are required.

The application of geographic methods to understanding trade areas dates to the pioneering work of geographer William Applebaum during the 1930s. Most of what is presented in this chapter is a technological variation on the basic principles introduced by Applebaum (1968) Applebaum's principles remain valid today. But when applying Applebaum's methods, geographic information systems (GIS) increase productivity and accuracy of measurement. During the 1960s, Applebaum began to publish the methods he had developed several decades earlier as a practitioner-academic (Applebaum 1965a; Applebaum and Cohen 1960; see also Hoyt 1969). Among these methods were the *customer spotting technique* and the *analog method*. Both of these methods are explained below.

Applebaum (see Applebaum and Cohen 1960) introduced factors for evaluating store sites and identified how location quality contributed to store rents. Among the elements considered were site accessibility, trade-area population, competition, economic stability of the region, trade-area penetration, store size and store function, building costs, operating costs, insurance, taxes, maintenance, and so forth. After identifying the important elements, Applebaum proposed a checklist for evaluating store sites and store rents. This checklist established the importance of the elements on the list for the evaluation of a site and land value. Next, Applebaum applied his analog method, customer-spotting technique, and trade-area delineation methods in the calculation of three important elements on the list: store trade areas, market penetration, and potential sales (Applebaum 1966).

Applebaum used a combination of proprietary data collected by the firm (such as customers' locations) and secondary data including demographic information and the location of competitive facilities. He used population data from the U.S. Census, license plate data from state agencies, and information provided by the U.S. Census of Retail Trade. This method was introduced was in the first third of the twentieth century, well before the desktop computer.

Applebaum's methods enable analysts to consider the issues of unserved trade areas and overlapping trade areas, including the consequences of cannibalization that result from overlapping trade areas. In the process of revealing the origins of customers, customer spotting also reveals the *range*, which is the distance customers will travel to patronize a store. (Berry and Garrison 1958; King 1984). Customer spotting also reveals the *distance decay* of customers—how the density of customers falls off with increasing distance from the real estate project, such as a retail site. Knowing the existing trade areas, knowing how the addition of new real estate projects will likely affect trade areas, and knowing the expected spatial behavior of customers are all necessary information for an analysis of the potential success of a real estate project.

The Customer Spotting Method and Trade Area Categorization

Applebaum obtained customer addresses by conducting in-store customer surveys and recording the license plates of customers' automobiles. Contemporary geographic technology offers many ways of obtaining customers' addresses that are much less time consuming than Applebaum had to contend with. These include bank data (ATM records, check records), credit card data, and point-of-sale data such as by ZIP code (see Thrall 2001).[2]

Information about origins of customers can be used to accurately delineate trade areas. But what exactly is a trade area? Applebaum (1966) proposed three types of trade area zones: primary, secondary, and tertiary.

The *primary trade area* is the geographic core from which a real estate project will get the majority of its business. Applebaum (1966) used supermarket data for metropolitan areas to reveal that the core trade area accounted for 60–70 percent of the real estate project's customers. Analysts today often define the primary trade area as having 80 percent of the customers. The *secondary trade area* contains the next highest percentage of the real estate project's customers. It accounts for 15–25 percent of customers. The *tertiary trade area* is geographically situated at the market-area fringe and accounts for the residual, or less than 15 percent of customers. The tertiary trade area can include unaccounted-for customers, sporadic customers, and out-of-town customers.

Applebaum (1965b) also considered the role of agglomeration on trade areas (see also Hoyt 1958). *Agglomeration* occurs when mutually beneficial businesses locate in close geographic proximity to each other. Contemporary regional malls and power centers can often out-compete stand-alone stores with superior locations because of the benefits agglomeration brings. To the customer, the benefits include the ease of multiple-purpose shopping and familiarity with the location. Applebaum performed a case study comparing the trade area for a discount food supermarket located adjacent to a general discount merchandise store with the trade area of a comparably isolated discount food supermarket. The supermarket located with the merchandise store was found to have more drawing power than the spatially unassociated supermarket. Thus, Applebaum demonstrated that geographic strategic alliances allow real estate projects to increase their trade areas and the percentage of population that become patrons (market share) from within each trade area zone. One of the uses of gravity and spatial interaction models, discussed below, is the calculation of trade areas with and without the benefits of agglomeration.

The Analog Method

Applebaum developed and began to apply his analog technique (Applebaum 1968) of sales forecasting in 1932 (Rogers and Green 1978). This analog procedure was the first systematic retail forecasting system founded on empirical data (Ghosh and McLafferty 1987). Applebaum (1966) proposed the analog technique be used when the store characteristics, market factors, consumer shopping behavior, and sales all have supporting data measurements. The resulting tabulation of data becomes a record of experience, an analog useful as a benchmark for future reference.

The analog model proposition is that a set of stores located in similar environments will reveal patterns of customer patronage and retail sales that can be used to anticipate the same for other stores located in similar, or analogous, environments. Existing stores are then the analogs against which other stores can be compared. "Other stores" can be existing stores or even proposed stores considered for the future. This method then introduces the ability to calculate geographic what-if scenarios. What if a store is built at a specific location? What will be its anticipated trade area, market penetration, and retail sales?[3]

Applebaum's analog method has strengths and weaknesses. The strengths include:

- Performance can be adequately measured by market share and per capita sales penetration (Buxton 1993).
- Many explanatory variables can be included in the assessment.
- Prospective sites for development can be analyzed, compared, and ranked according to their expected performance.

The weaknesses of Applebaum's analog method include:

- New markets with unusual demographic and competitive profiles may not match any markets in the analog base (Buxton 1993). As markets differ from one another and as markets become more competitive, the value of the analog method decreases (Goldstucker et al. 1978).
- As the database of customer spotting grows larger, the ability of the analyst to organize and retain data and relationships becomes impaired, perhaps leading the analyst to select analog groupings in a faulty, judgmental and arbitrary manner (Rogers and Green 1978). The stores chosen by the analyst for use as analogs can be highly subjective, and bias is easily entered into the model.

- With regard to existing store analysis, the analog method provides no quantitative expression or test of the possible relationship between the extent of a trade area or the level of market share and population or competitive conditions (Rogers and Green 1978).

Predicting Revenues with the Analog Method

The following section presents an analog-based regression procedure for estimating revenues of a real estate project. A retail development is used in the example, but revenue forecasts of most real estate projects will follow similar methodology. First, background is provided on the statistical procedure of regression analysis.

Regression Analysis Overview

Regression is fundamentally an analysis of relationships among variables (Chatterjee and Price 1991). With regression, we determine the effects of one or more explanatory, or independent, variables on a dependent variable. A simple univariate regression equation, with one explanatory variable, can be written as:

$$Y = b_0 + b_1 X, \tag{4.1}$$

where b_0 and b_1 are the regression coefficients that the statistical procedure calculates based on the data sample for the study. The coefficient b_0 is also the vertical intercept, X is the explanatory or independent variable, and Y is the dependent variable (i.e., the variable being explained by the regression analysis).

Equation 4.1 is structurally identical to the general equation for a straight line. The purpose of a regression equation is to determine how much of the variation in Y is explained by X. Regression analysis essentially estimates the value of the intercept and the slope of that straight line and provides measures of confidence for those estimates.

Now regression analysis is introduced into Applebaum's methodology. Y may be defined as weekly sales of a commercial real estate project, measured in thousands of dollars. Say that the analyst has determined the geographic extent of the primary trade area and the underlying geodemographic characteristics of the primary trade area. Say that the primary trade area is calculated as a circle with a radius of five miles. Among the geodemographic characteristics of the primary trade area is median per capita income of the population residing within the primary trade area, namely, X, also measured in thousands of dollars. For every $1000 increase in per capita income of the population within the primary trade area, weekly sales are expected to increase by $500 (0.50 × $1,000). Say that the analyst has estimated the vertical or Y-intercept to be 9.2, measured in thousands of dollars, meaning that $9200 of weekly sales are independent of the income characteristics of the primary trade area. In other words, regardless of the income of the population within the primary trade area, minimum sales of $9200 per week are expected. Substituting these numbers into equation 4.1 yields:

$$Y = (9.2) + (0.50) X. \tag{4.2}$$

Next, consider a situation where an analyst is evaluating the feasibility of a prospective site. Using the above regression model as an Applebaum-like analog, the analyst can estimate the expected sales at the proposed location, given the median per-capita income of residents within the five-mile radial primary trade area. Based on the experiences of the same kind of retail stores at other locations, expected weekly sales are calculated using equation 4.2. Let the median per-capita income of residents within the five-mile radial primary trade area be $20,000. Substituting that figure into equation 4.2 yields:

$$Y = (9.2) + [(0.50)\ (20,000)] \qquad (4.3)$$
$$= \$19,200.$$

Therefore, $19,200 is projected to be the weekly sales for the proposed commercial real estate development.

Suppose the real estate market analyst knows that the median age of the population within the primary trade area is also important. The above regression equations can be modified to include more than one explanatory variable on the right-hand side of the equation. When two or more explanatory variables are included in the model, the regression equation, known as a multiple regression or multivariate regression, can be expressed as:

$$Y = b_0 + b_1 X_2 + \ldots + b_n X_n. \qquad (4.4)$$

where $b_0 \ldots b_n$ are regression coefficients and $X_1 \ldots X_n$ are the independent variables (for more examples of multiple regression equations, see Davies and Rogers 1984; Drummey 1984; Rogers and Green 1978).

The above example can be extended to a situation where experience suggests that weekly sales depend on a variety of geodemographic characteristics of the primary trade area, including median income, population density, traffic count, and median age. The multiple regression equation would then include these variables, thereby providing a more accurate estimate of weekly sales, and simultaneously validate the importance of these variables as relevant descriptive characteristics of the trade area. Examples of multiple regression are provided elsewhere in this book.

One indicator of the explanatory power of the regression model is the coefficient of determination, commonly referred to as R^2. The R^2 is a measurement of how much variation of the dependent variable is explained collectively by the independent variables. It is the most frequently used measure of goodness of fit. The numerical range for the R^2 is between 0.0 and 1.0, where a value of 0.0 indicates that no variation in the dependent variable has been explained by the independent variables, and a value of 1.0 indicates that 100 percent of the variation in the dependent variable has been explained by the independent variables (i.e., a perfect model). If $R^2 = 0.74$, then 74% of the variation in the dependent variables had been accounted for by the independent variables.

A number of problems can be encountered when using regression analysis; chief among these for the real estate market analyst is multicollinearity. Multicollinearity occurs when two or more independent variables are highly correlated, meaning that the two variables explain essentially the same thing. For example, income generally increases with age. Including both income and age as independent variables may therefore introduce multicollinear bias into the model, resulting in unreliable estimates.

In addition to estimating the parameters of the regression equation, the analyst usually constructs a correlation matrix of the independent explanatory variables. If the independent variables produce correlation values greater than 0.5 or 0.6, then the results of the regression analysis may be invalid, and the analyst should omit the offending variables from the equation. Another indication that multicollinearity is biasing the calibration of the regression equation is the statistical insignificance (t values) of the individual regression coefficients (B values), even though there may be a high degree of conformity for the model as a whole, measured by a high R^2 (for further discussion on this topic, see Ghosh and McLafferty 1987).

Another problem of regression analysis that the real estate market analyst must deal with is determining the relative importance of the independent variables. That is, which among a pool of variables hypothesized to be influential in determining revenues is the most important and statistically relevant? Multicollinearity might preclude the analyst from using all the variables that have been hypothesized to be important in the regression equation. To determine relative importance of variables, many real estate market analysts use a variation on regression analysis known as forward stepwise regression.

Forward stepwise regression allows for the sequential introduction of independent variables into a regression equation. The independent variable that explains the most variation in the dependent variable is introduced in the equation. Then the next most important independent variable is introduced. This is repeated until either no variables remain, or until no variables are left that if introduced into the equation will be statistically significant (for further discussion and criticism, see Thrall 1986, 1988). This procedure generally avoids the problem of multicollinearity. Forward stepwise regression can also increase the R^2 of the model. In general, the greater the R^2 of the model, the more accurate the predictability of the model. Many analysts also use forward stepwise regression to reveal which variables the data reveal are important.

Regression analysis is important to real estate market analysis because it allows for a number of identifiable site and situation characteristics to be successfully isolated, quantified, and assessed. However, the use of regression models for commercial real estate site analysis is based on two fundamental assumptions. The first is that the project's performance will be significantly affected by the characteristics of the project's location, the geodemographic composition of the primary trade area, the level of competition, and store characteristics. The second assumption is that the characteristics hypothesized to be important can be measured and isolated by systematic analysis (see Ghosh and McLafferty 1987; Jones and Mock 1984).

Real estate market analysts include variables hypothesized to measure the geographic trade area of the retail store. The trade area may be measured using one of the three methods explained below in this chapter. Included in the regression equation are geodemographic variables of the customers and of the trade area. Also included in the model are descriptive characteristics of the real estate project itself, such as the project's square footage, number of employees, the length of the storefront, number of checkouts, range of merchandise, parking facilities, and amenities of the location (for further discussion, see Simmons 1984).

It is not only the list of variables included in the model that is important, but how the variables are mathematically included. For example, multicollinearity is a problem for retail store models when the dependent variable is sales and the independent variable is the square footage of the store (Lea 1989). Transforming the dependent variable into sales per square foot can reduce the problem of multicollinearity. An added benefit of this transformation is that as the square footage of the store increases, the proportional increase in sales can be measured.

Residuals are another important consideration of regression analysis, and can be a source of error. Residuals are calculated as the difference between the value observed for the dependent variable and the value that the calibrated equation predicts for that observation. Geographers have mapped regression residuals for more than 30 years (e.g., see Lea 1989). Residual mapping can be used to identify variables that should be either included or excluded from the model. Residuals can also be used to identify which retail locations exceed or fall below expected sales forecasts (see Ritchey 1984). A company with many retail stores could use residual information in its long-term strategic plans (Lord and Lynds 1981). If a map of residuals reveals a regular geographic pattern, then a problem known as *spatial autocorrelation* may exist. Spatial autocorrelation is similar to multicollinearity and occurs when the observed values of the independent variables depend on their geographic location, which in turn affects the values of observations at neighboring locations (see Odland 1988).

Regression Analysis and the Analog Procedure

Regression analysis has become a standard tool of real estate market analysts. Applebaum's original analog methodology, however, did not include regression analysis. Lord and Lynds (1981) used regression techniques to identify location variables affecting store sales. Davies (1977) included locational variables in a regression model that estimated revenues for new locations for commercial real estate.

Cottrell (1973) evaluated supermarket performance using both regression analysis and a variation of Applebaum's method. Cottrell's model separated supermarket performance into external and internal environmental factors. The residual of the attributes of the store's location was interpreted as a measure of the store manager's performance. Cottrell's model included social, demographic, and economic descriptions of the store's external environment; the presence of similar retail stores; the configuration of the individual retail store, including hours of operation and price range; and the level of managerial effort (Cottrell 1973; see also Hise et al. 1983).

Ghosh and McLafferty (1987) also applied regression models to identify stores performing above or below expected levels. They too believed that the residual is attributable to managerial effects. Thus, multiple regression models have incorporated both location and nonlocation variables in the assessment of store performance.

A decision must be made about which stores to include in the sample data set used to calibrate the regression model. For example, say there is a restaurant chain with sites in the central business district and at urban-perimeter shopping malls.[4] It might be anticipated that because the environments are distinctly different, the two types of locations will exhibit dramatically different behaviors. In other words, the retail chain will need two models, one for city center, and another for suburban locations. The observations from the categories would not then be pooled.[5] Along this line of reasoning, Davies (1973) discovered that the same variables included in a regression equation predicted sales for stores located on a corner differently from for stores located in the middle of a block. The analyst must therefore carefully consider the parallel issues of stability of calculated parameters of the regression model and how the data are collected and defined.

Performance may also depend on time. A newly opened store may not experience the same level of sales as a store that is three to five years old. For this reason, Lea (1989) proposed that data be included in the calibration of the regression model only for stores that have reached a stage of maturity in their operations.

Real estate market analysts also use other related methods, including *analog regression* (not to be confused with Applebaum's analog procedure). The performance of a store is usually the dependent variable when applying regression techniques to retail location analysis. But in analog regression, the market share of submarkets of the store's trade area is used as the dependent variable. Geodemographic character-istics of the submarkets are used as the independent variables.[6] When Ghosh and McLafferty (1987) applied analog regression to retail site analysis, they used as ex-planatory variables distance separating a submarket from the store, store size, and the size of the nearest competitor. This selection of variables is based on concepts bor-rowed from the gravity model approach, which is explained below.

An advantage of analog regression is that multiple submarkets can be defined for a single retail location, resulting in a sizeable increase in the number of observations available to the analyst (Davies and Rogers 1984). Analog regression can be partic-ularly useful to retail chains with a small number of stores. Customer spotting can used to calculate the market share of each submarket for a store (Ghosh and Mc-Lafferty 1987). The total sales forecast for a store is then calculated as the sum of the forecasts for each submarket.

What is the role of GIS in model estimation and regression? Just as regression analysis was a natural next step to implement Applebaum's analog methodology, GIS is the natural next step after the concepts of regression analysis are internalized. In other words, we first conceptualize the problem, identify or develop the proper pro-cedures to achieve a solution, and streamline the procedures so that they are more routine and efficient. GIS is the data organization engine and the vehicle for visualizing the results on a map. GIS technology is the vehicle for making the procedure more efficient, accurate, and accessible to analysts, and it increases the productivity of the analyst (for further discussion of regression models, see Davies 1977, Drummey 1984; Ingene 1984; Ingene and Yu 1981; Jones and Mock 1984; Lee and Koutsopoulos 1976; Simmons 1984; Wilson 1984.)

Geographically Delineating a Trade Area

In the following section, three general methods for geographically delineating a trade area are presented. Geographic boundaries of trade areas are derived using rules of thumb, gravity and spatial interaction models, and customer surface modeling. The ge-ographic boundary of a trade area establishes which data are included in the analysis.

Market Area Delineation via Rules of Thumb

The two rules of thumb that are most frequently used (with varying levels of success) for deriving the geographic extent of a market area are the radial distance approach and the drive-time approach. These two approaches for defining the geographic extent of market area have already been described in chapter 2. Figure 2.9 illustrates radial distance bands around a central location, and figure 2.10 illustrates a driving-time distance band (Thrall 1999).

The *radial distance approach*, also known as rings, has been used more for analytic convenience than for accuracy of representation of the trade area. Circular trade areas arose during the time when geodemographic analysis was calculated by hand, before desktop computers allowed for quick calculations of more complex shapes. A circle is easy to draw on a map. Applebaum's customer spotting method was used to reveal the locations of customers, and a circle was drawn to inscribe the chosen percentage of customers, thereby forming the primary trade area. The circular primary trade area was superimposed on a map containing boundaries of five-digit ZIP codes or census tracts. This then identified which ZIP codes or census tracts to include in the trade area. Geodemographic data, like those in table 2.1, have been most commonly available by ZIP code or census tract, and are obtained from private data vendors or government. The analyst then uses the resulting information to calculate expected sales or revenues, following the procedures outlined above.

Repetitive application of Applebaum's customer spotting method gives rise to rules-of-thumb concerning the appropriate radius to represent the primary trade area. These rules of thumb are specific to various real estate product niches, such as restaurants or regional malls, or specific to particular chains within these niches. For example, a restaurant chain might use 5 or 6 radial miles to delineate its primary trade area based on experience that between 70 and 80 percent of its customers usually originate within that distance.

Circular trade areas are expedient to calculate, but they can be the cause of unnecessary error in the trade area analysis. A circle may have no meaningful correspondence with other geographic patterns, such as transportation networks, which influence customer geographic behavior. Also, the geometry of an expanding radius for a circle may dictate that in order to include an additional five percent of customers located in the northeast, an equal area will need to be added to the trade area in other directions, even though no prospective customers are located there. Instead, an irregular polygon can be used to inscribe those additional customers without necessitating that other areas, perhaps devoid of known customers, be included as well.

A variety of algorithms have been used to derive irregularly shaped polygons for use as trade area boundaries. These GIS based algorithms create smoothed ellipsoidal polygons using filters, weights, and other controls. Customer and store locations are mapped. The customer data records can be weighted individually so that each customer location is not equal. Individual customer generated sales is an example of a weight, as is number of purchases. Pie shaped wedges are generated, rather than a circular trade area polygon. The algorithm allows each wedge to expand its radius outwards, subject to each incremental expansion adding to the whole trade area a minimum number of customers or their weighted value. Or, a wedge that is able to add the most in the next increment outwards is increased in radius, then the algorithm proceeds to another round of possible expansion. The algorithms in common use today allow the user to specify the number of pie wedges surrounding a store, and the percent of customers to include in the final trade area derivation. Most algorithms allow for smoothing of the perimeter of the trade area polygon. Figure 4.1 shows a trade area derived using such an algorithm. The algorithm defined eight wedges, and captured

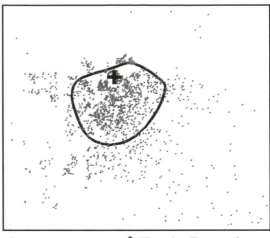

Figure 4.1 Irregular polygon trade area generated by geoVue Inc.'s MarketSite software

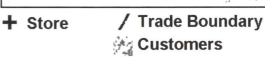

80% of the total customers within the trade area polygon. Note that the resulting trade area is not circular; rather, the store is in the upper portion of the trade area which expands more to the south than to the north (Thrall and Casey, 2001).

The transportation network strongly influences travel behavior of prospective customers. Desktop GIS software can calculate trade areas based upon a specified time to drive between the origin of the real estate project and all possible destinations as in figure 2.10. Also, the irregular polygon of figure 4.1 could have been calculated using transportation distance or time algorithms to identify the proximity of customers to the store. Market areas based upon drive distance or drive time are generally considered to produce a more accurate trade area; GIS allows for the efficient derivation of trade areas based upon complex spatial algorithms. The sample geodemographic report shown in table 2.1 includes demographic profiles of people located within ten, fifteen, and twenty minute estimated driving times from a central place.

A combination of Applebaum's customer spotting method and drive time estimations can be used as a rule of thumb to inscribe the primary trade area. GIS software can calculate the required trade area whose boundary requires drive time to be no greater than some specified amount and that inscribes the required percentage of customers. Repetitive calculation of such drive time trade areas will give rise to rules of thumb on travel time that might be used to estimate the primary trade area for real estate projects before they have been developed. Likewise, how much time the typical customer is willing to allocate to drive to a destination with a given set of characteristics can also be estimated.

Market Area Delineation via Gravity and
Spatial Interaction Models

The gravity model can be used to derive the primary market area of a real estate development. The gravity model is a procedure whereby the market share of population resident at one location is calculated to interact at another location. Trade area delineation with a gravity model largely depends on characteristics of the real estate development, as opposed to characteristics of the population as with Applebaum's customer spotting model. Calibration of a gravity model can be improved, however, using customer spotting and then calculating customers' friction of distance. Gravity models are particularly useful when there is insufficient information about the geographic behavior of customers, while at the same time the developer can observe the change in trade area as characteristics of the real estate project changes.

Origins of the National Model   More than a century ago, Newtonian physics was applied to measure social phenomena, including market areas.[7] H. C. Carey (1858), with the publication of his book *Principles of Social Science*, revolutionized the application of Newtonian measurements of gravity to analyze location, including market areas. Twenty-seven years later, E. G. Ravenstein (1889) used Newton's gravity model to measure human migration. And forty-four years after that, W. J. Reilly (1929) used the gravity model to calculate the market area surrounding a shopping center. Reilly's rediscovery of Newton's gravity model is now associated with what is perhaps the most familiar economic geography application in real estate market analysis and marketing: *Reilly's law*.

The literature published after Reilly retained the fundamental theme of the gravity model but resurfaced under a variety of names, including Zipf's minimum effort model (Zipf 1949); Huff's model of consumer behavior (Huff 1959, 1963, 1964);[8] Stouffer's intervening opportunity model (Stouffer 1960); and Wilson's entropy model (Wilson 1967). The term *spatial interaction model* is normally preferred by academics in recognition that new formulations are in no way dependent on Newtonian physics, but instead follow from clear and precise geographic reasoning. "Spatial interaction model" more clearly describes the purpose of the model, which is to calculate the movement or interaction between spatial phenomena. However, the term "gravity model" has been in common usage for more than 150 years, and for that reason is most commonly used among practitioners (for additional discussion, see Berry and Parr 1988; Ghosh and McLafferty 1987; Haynes and Fotheringham, 1984).

The Gravity Model   Figure 4.2 shows the basic gravity model relationship. The gravity model is based on Newton's hypothesis that interaction between the two objects is directly proportional to the mass of the objects and inversely proportional to the distance between the two objects. So, as distance increases (all other things being equal), the interaction between the two objects decreases. Conversely, as mass increases, so does the interaction between the two objects.

In figure 4.2, the populations of city x and city q are 2,000,000, and the population of city z is 1,000,000. The potential pairs of interactions between cities x and z are then 2,000,000 × 1,000,000. Because city q has twice the population of city z, the interaction between cities x and q would be twice that of cities x and z—namely, 2,000,000 × 2,000,000. The interaction between two places is therefore calculated as a multiplicative function of the masses, where population count serves as the measure of mass.

City y and city q have the same population, but the distance from city y to city x is twice the distance from city q to city x. The interaction between cities x and y is therefore expected to be less than that between cities x and q.

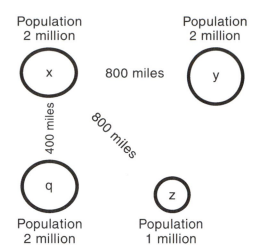

Figure 4.2 Basic gravity model relationship (after Haynes and Fotheringham 1984; see also Thrall and del Valle 1997a).

From the information in figure 4.2, we can calculate the expected interaction between places where interaction increases as a multiple of the masses involved and decreases as distance separates the places:

$$T_{ij} = P_i P_j / d_{ij} \qquad (4.5)$$

In equation 4.5, T_{ij} is a measure of the total interaction between places i and j, with population counts, respectively, P_i and P_j. Distance separating the places is d_{ij}. By how much does distance change interaction? Distance in equation 4.5 is assumed to reduce interaction at a rate of d_{ij}^{-1}.[9]

Nineteenth-century geographers literally interpreted Newton's formulations, which led them to believe that interaction must decline at a rate equal to the square of the distance. That error has persisted down to today; it is still common to measure the *friction of distance* as the inverse square of distance between two places. However, not all phenomena have the same friction of distance, so we normally represent the friction of distance as $d_{ii}^{-\beta}$. Variations on Applebaum's methods described above, and in-store surveys, are used to calculate appropriate measurements for the friction of distance, β.

As it has been reasoned above that β will vary between types of real estate projects, including different chains within the same product niche, then we should also consider that the population terms in equation 4.5 should also be weighted. Not all populations are equal when forecasting interaction: populations with higher incomes and greater education might be expected to have greater interaction with other places than have populations with lower incomes and lesser educations. Indeed, when measuring the mass of two places, specific characteristics used to define a target demographic population might be used, thereby reducing the total count of persons to more accurately reflect the relevant population. Target demographic populations are often grouped by their LSP (lifestyle segmentation profile), as discussed in chapter 2 and applied in subsequent chapters. Therefore, a weight is applied to the population of place i equal to the value of the exponent λ, and a weight equal to the exponent α will be applied to the population of place j. The larger the exponents α and λ the greater the attraction or propensity of the population to interact.

As a final modification to equation 4.5, a constant of proportionality, k, is included so that the numerical calculations generated from the measurement can be interpreted with a known scale or unit of measurement.

Including the above modifications to equation 4.5, the spatial interaction equation becomes:

$$T_{ij} = k\, P_i^{\alpha} P_j^{\lambda} / d_{ii}^{\beta}. \tag{4.6}$$

Variations of equation 4.6 are among the most commonly applied geography formulations in real estate market analysis. Formulations like equation 4.6 allow GIS technology to be more than a descriptive depiction of an urban landscape; instead, GIS coupled with appropriate geographic reasoning and modeling can become an explanatory and predictive methodology.

The Reilly Model Reilly's use of equations 4.5 and 4.6 became so popular among academics and business geography practitioners early in the twentieth century that the formulation came to be known as Reilly's Law (see Converse 1949, Craig et al. 1984, Okoruwa et al. 1994). It should be recognized that the formulations Reilly used are not laws, nor was he the first to apply Newtonian physics to analysis of the human-built environment.

In 1931, Reilly wrote that a city will "attract retail trade from any intermediate city or town in the vicinity of the breaking point, approximately in direct proportion to the populations of the two cities and in inverse proportion to the square of the distance from these two cities to the intermediate town" (Reilly 1931, p. 3; see also Reilly 1929). Reilly's formulation is a variation on equation 4.6, where the friction of distance is 2, the exponential weights for the population α and λ are both equal to 1.0, and the constant of proportionality, k, is also 1.0. The Reilly formulation for the attraction, A, to city i by prospective customers in city j to city i's goods and services is:

$$A_i = P_i / d_j^2. \tag{4.7}$$

Using equation 4.7, Reilly examined the competition between retail centers for a market; developed a method for estimating the flow of consumers or expenditures from a market area hinterland to competing market centers; and developed a method for estimating a market area boundary. The first of Reilly's steps was to calculate the proportion of market share to be received at each market center. The attraction of each market center from some location on a map is measured as the population of the market center divided by the distance separating the market center and a chosen location:

$$A_i = P_i / d_{ix}^2, \tag{4.8}$$

$$A_j = P_j / d_{jx}^2. \tag{4.9}$$

The proportion share received at a market center is calculated as $A_i{:}A_j$. The following example makes clear the meaning of the ratio $A_i{:}A_j$.

Table 4.1. Measurements for the spatial interaction example

From	Distance to			Population
	Dallas	Wichita Falls	Oklahoma City	
Dallas		100	200	2,500,000
Wichita Falls	100		140	350,000
Oklahoma City	200	140		1,000,000

Based on Haynes and Fotheringham (1984).

The necessary numerical values used in the calculations are presented in table 4.1. In this illustration, all that is required are distances between the places and their populations. It is assumed that the market is composed of only three cities. An analyst applying the Reilly procedure would first identify all those market centers that are expected to be competitive with the target market center. The analyst would repeat the procedure for each pair of competing and target market centers. Using the information from table 4.1 and the above equations, the market attraction from Wichita Falls to Dallas and Oklahoma City, respectively, is

$$A_i = P_i/d_{ix}^2 = 2,5000,000/100^2 = 250,$$
$$A_j = P_j/d_{jx}^2 = 1,000,000/140^2 = 51.02.$$

The ratio of the attraction for each market center is then

$$A_i:A_j = 250/51.02 = 83.05\%$$

The result indicates that out of all of the trade flow, or expenditures, emanating from Wichita Falls out to Dallas or Oklahoma City, Dallas will receive 83.05% of that trade flow. The remaining 16.95% (100% − 83.05%) will go to Oklahoma City.

The second step in Reilly's formulation is to calculate the area over which a particular trade center is dominant. Consider that at some location on the map, the attraction of going to one city or the other will be equal. A location where the difference between the two competing market centers is inconsequential to a household is one measure of a market boundary. In other words, determine the location where $A_i = A_j$, or

$$P_i/d_{ix}^2 = P_j/d_{jx}^2 \tag{4.10}$$

In Euclidean geometry, it is well known that the whole is equal to the sum of its parts. Therefore, the distance $d_{ix} = d_{ij} − d_{jx}$ when cities i and j and the market-boundary breaking point x lie on a straight line. Using this fact, we can rewrite equation 4.10 as

$$d_{ix} = d_{ij}/[1 + (P_j/P_i)^{1/2}]. \tag{4.11}$$

The general form of Reilly's model when friction of distance is not restricted to the square of the distance is

$$d_{ix} = d_{ij}/[1 + (P_j/P_i)^{1/\beta}]. \tag{4.12}$$

Figure 4.3 Hypothetical Reilly market area problem (after Haynes and Fotheringham 1984; see also Thrall and del Valle 1997).

Substituting the values from table 4.1 into equation 4.12, we can solve the breaking point between Dallas and Oklahoma City, which is

$$d_{ix} = 200/[1 + (1{,}000{,}000/2{,}500{,}000)^{1/2}]$$
$$= 122.5.$$

Therefore, as a first estimation assuming that the friction of distance is measured as the square of the distance, it is calculated that the distance over which Dallas dominates is 122.5 miles. The distance over which Oklahoma City is calculated to be dominant is the difference between the total distance separating Oklahoma City and Dallas (200 miles) and the distance over which Dallas dominates (122.5 miles): 77.5 miles toward Dallas. Figure 4.3 shows the results of the calculations.

Wichita Falls is within Dallas' market boundary, but that does not mean that Dallas will capture 100 percent of the Wichita Falls market. Dallas will receive 83.05 percent of Wichita Falls's patronage. Figure 4.4 shows the market share calculated using the numerical values from table 4.1 and the formulations based on equations 4.8 and 4.9. Dallas, being the larger market center and having a greater gravity, or mass, has a larger market area of dominance.

When an analyst creates a GIS application, the procedure is repeated for each

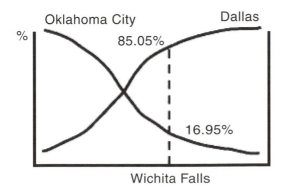

Figure 4.4 Market share calculated from values in table 4.1.

competitive real estate project, be it a regional mall or a restaurant. The analyst using GIS would then evaluate the demographic composition within the derived trade area polygon and surrounding areas as explained above.

Complicating Factors   The Reilly model is attractive to practitioners because of its simplicity and because the analysis addresses important questions. However, Reilly's model does have a number of limitations. Market attraction is actually much more complicated than ratios of mass to distance. The taste preferences of the population may change and favor one market center over the other. There was a time when households in Toronto, Ontario, flowed in great numbers to Buffalo, New York, for shopping and entertainment. However, because of the dramatic decline in the quality of the urban built and natural environment of the Buffalo region, the direction of flow has reversed to Toronto. The dimension of time and changing urban amenities are not incorporated into this presentation of the Reilly model, nor are the components of household income and socioeconomic barriers.

The issue of types of goods also exists. Households may be willing to travel great distances to save small percentages on high-cost durable goods such as high-end television sets, but they may not be willing to travel the same distance to save high percentages on low-cost goods consumed on a daily basis. These factors have not been taken into account in the above example; therefore, the analyst should proceed with caution when applying the Reilly formulation.

Market Area Delineation via Customer Surface Modeling

Applebaum's customer spotting method and mapping the density of customers can reveal much about a trade area (Donthu et al. 1989). The mapping can indicate the general location of the boundary, as well as indicate the presence of significant geographic submarkets. This section explains the kernel method for mapping the density of customers and for revealing the geographic boundaries of the market area.

Interpreting the Kernel   Visualizing the density of customers is one of the best means of revealing the geographic limits of trade areas. The *kernel method* is currently the most commonly applied method to map customer density. During the 1980s, the kernel method became the generally accepted procedure for density estimation, and recent studies continue to support the use of this technique (Donthu et al. 1989; Rust 1986; Sabel 1998; Williamson et al. 1998).

Figure 4.5 is a map produced using the kernel method. More darkly shaded areas represent locations calculated as having a greater density of customers than lighter-shaded areas. One darkly shaded submarket is near the store location. The store location is indicated by a triangle. There are also other lesser local peaks of customer density. The peak density near the store might be a manifestation of high store and customer proximity. However, other explanation is required for the other lesser, localized peaks.

The peaks of customer density that are not surrounding the store might result from the type of people who have clustered within those neighborhoods. Real estate market analysts often describe people by their LSPs (see Thrall and McMullen 2000a). People with similar LSPs tend to live in close proximity to one another and thereby form

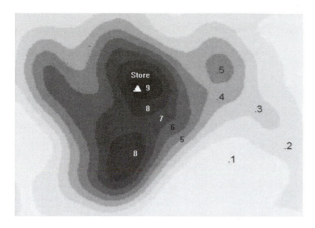

Figure 4.5 Map produced using
the kernel method (from Thrall
and McMullin 2000b)

neighborhoods of people sharing similar lifestyles and demographic profiles. People
that share similar lifestyles and demographic profiles are highly correlated in their
consumption patterns as well. Therefore, the market analyst might consider the geographic submarkets revealed by the darker-shaded areas as neighborhoods containing
the greatest number of prospective future customers. Further, by deriving the LSPs of
people within the peak geographic submarkets, the market analyst will have calculated
a profile of the target demographic population (Birkin 1995). When searching for new
trade areas to penetrate, the business geographer often begins by constructing maps
for the new region of the density of people that share the targeted LSP.

Describing the Kernel Function   The kernel is a three-dimensional function that
moves across a map. The kernel has a radius referred to as a *bandwidth*. The kernel
is said to visit each of mapped customer locations and weighs the area surrounding
the customer proportionally to its distance to other customers. The kernel value for
each customer within the study region is thereby calculated. A smoothed surface is
then produced of those values (Sabel 1998). The kernel function is a measurement of
the average proximity of customers to one another. The real estate project, such as a
retail store, is often situated somewhere inside the study region, but that is not necessarily the case. Because of Internet commerce and mail order catalogs, there may
not be a physical facility within the trade area. Still, the geographic trade area remains
relevant. The kernel uses the location of the customers to reveal the trade area. In
contrast, as described above, the gravity and spatial interaction model relies more on
the location and characteristics of the store (Huff 1964; Huff and Rust 1984; Thrall
and del Valle 1997).

A map like that of figure 4.5 might be used to advise on the location of service
facilities and warehouses, as well as new retail outlet facilities (see Goldstucker et al.
1978). Kernel maps might also be used to identify neighborhoods to target for "customer mining."

Applying the Kernel Method   The first step when applying the kernel method is to
create a map showing the geographic distribution of customers, as in Applebaum's

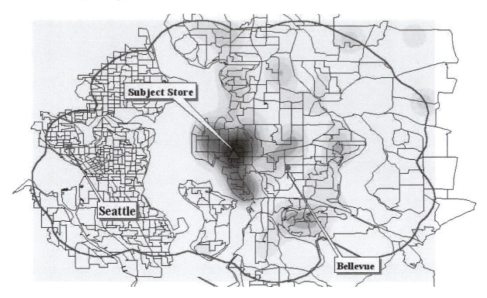

Figure 4.6 GIS derived trade area boundary and customer expenditure surface (from Thrall and McMullin 2000b).

customer spotting method. This might be created using an in-store customer survey. GIS technology allows for the derivation of customers' latitude and longitude coordinates based on their addresses. Other GIS technology can assign a derived LSP to each customer record (see Thrall and McMullin 2000a).

The second step is to construct the boundary of the primary trade area so that a specified percentage of customers is included within the boundary, such as 80 percent. Various GIS software programs include commands to perform this operation, as shown in figure 4.6. The resulting boundary created is an irregular polygon. See also figure 4.1 and related discussion.

The third step is to create a smoothed geographically interpolated surface. Various GIS software programs include commands to perform this operation. The result is a map like those of figures 4.5 and 4.6.

Trade area analysis is not limited to displaying customers and their density surface on a map. Customers can be weighted by a variety of characteristics, most important of which is generally customer expenditure. In figure 4.6, darker areas reveal concentrations of high sales, while lighter shades reflect lower sales. Mapping the density of customers and customers' expenditures can reveal where the core of the market area is, as well as the existence of significant geographic submarkets. This information is useful for honing in geographic markets and evaluating what kind of customers inhabit the most important geographic submarkets.

Now that the trade area is understood, and can be derived, we proceed to estimate the supply of real estate product that is competitive with the subject project.

Step 2: Estimating Competitive Supply

Up through the late twentieth century, the calculation of the supply of competitive real estate was a time-consuming and arduous task.[10] An analyst might have gone to the county property records department and sorted through all real property records, noting those that could be construed as being competitive with a proposed project. Alternatively, the analyst might drive around the location, making notes of developed real estate projects. To measure the supply of competitive real estate projects that were at various phases of construction, including initial planning, the analyst would sort through files of building permits.

At the start of the twenty-first century, the geography technology industry was estimated at $50 billion per annum. Included as part of the geography technology industry are data vendors that have all but replaced the time-consuming and arduous task of pulling records from files. Today, the information that is needed in many markets is purchased and available either on CD-Rom or via the Internet. The productivity of business geographers consequently dramatically increased.

Several commercial data vendors provide information for estimating supply of real estate. Two of these products are discussed here to illustrate how competitive real estate supply at the present time, and in the pipeline, can be calculated. The data available via these products also provide a good first estimation of location and attributes of property peer groups (i.e., properties that contend as comparables for a subject property). F. W. Dodge, a subsidiary of McGraw Hill, provides the pipeline data commented on here, and Comps.Com, a subsidiary of the Co-Star Group, provides the commercial real estate stock data.

Real estate market analysts have a host of uses for data of this type. Uses include estimating the supply of real estate product at the present time and extending that estimate to the near future. Because these data are geographically enabled with latitude and longitude coordinates, they can be used to identify where the supply of real estate is and where it is increasing. In other words, submarket analysis can be performed. These data can also be used to identify the location and attributes of peer-group properties—properties that would be considered comparable. Although such data are highly useful to analysts, they do not conform to standards established by organizations such as the Appraisal Institute and should not be considered a replacement for data required for conventional appraisals.

Pipeline Data from F. W. Dodge

Pipeline data are part of the F. W. Dodge Market Analysis Group's package known as "Real Estate Decision Tools." The pipeline is real estate at various stages of construction from conceptualization to the awarding of a certificate of occupancy. What is in the pipeline can be used as a measure of the prospective increase in the supply of real estate. The analyst must be aware, however, that not all real estate projects are completed once they enter the pipeline. F. W. Dodge accommodates this by also including data on abandoned projects.

The pipeline data monitor five commercial real estate property types, in various stages of planning and construction, and are provided for 110 major markets. Alteration projects are included as well as new construction and additions. The data are updated quarterly. The commercial real estate property types included are apartment, hotel, office, retail, and warehouse. Dodge Pipeline includes all projects whose hard cost of construction is $1 million or above. The database includes both privately and publicly (i.e., government owned) owned projects.[11]

Real Estate Inventory Supply Data from Comps.Com

Comps.Com provides inventory data of existing commercial real estate. Comps.Com expects to provide coverage for between 70 and 100 of the largest U.S. urban markets. In 2000, data were available for 50 markets. Types of commercial real estate covered by Comps.Com are apartments, hotels, office buildings, retail buildings, warehouses, and shopping centers. Among these broad categories are various subtypes such as assisted living apartments for the elderly.[12] SQL-like queries allow the user to select from various automatically generated reports, including profiles on income, sales transaction history and valuation, contact, and mortgage information. The product includes a database of photographs and plat maps, which are available for viewing along with the other record information. Comps.Com data include latitude and longitude coordinates for each property. The data records of a query may be exported to a dBase .dbf file and then imported as a point data layer into desktop GIS software programs.

Comps.Com data can be used to estimate the current supply of commercial real estate. However, an analyst who relies only on current supply is ignoring that which is in the pipeline and that which may affect the future supply of the real estate product.

Using Pipeline and Real Estate Inventory

Both Comps.Com and F. W. Dodge data products provide latitude and longitude coordinates. Also, because addresses are provided for the properties, users may generate their own coordinates for the property. The geographic coordinates are a prerequisite for importing the data to a desktop GIS software program. Most desktop GIS software programs can import the data as a point layer. A GIS user may then perform queries on each of the data points and may perform other GIS operations as part of a larger analysis. Examples of such queries are:

- What is the supply of real estate by product category within five miles of a specified latitude-longitude coordinate, or address?
- What is the supply of real estate by product category within ten minutes driving time from a specified latitude-longitude coordinate, or address?
- What is the ratio of new supply of real estate by product category to existing supply of real estate product?
- What is the ratio of needed supply of real estate by product category (based on demand estimates) to present and future supply of product category? (This is commonly referred to as *absorption analysis*)

Real estate is information arbitrage (Thrall and Thrall 1991). Accuracy and timeliness of information are paramount to the real estate decision. These products are an example of how new technology is changing real estate market analysis and the way in which real estate decisions are made and demonstrate how the real estate decision is becoming dependent on information technology.

Economies of scale and multiple sales of data mean that the data from commercial vendors such as those described here and in Thrall 2001 are very affordable, especially considering the value added to the business decision and the cost of generating one's own equivalent data.

Step 3: Demand Estimation

Each category of real estate product requires different procedures for estimating demand. Procedures for demand estimation are therefore discussed in subsequent chapters on specific real estate product categories. However, common among all demand estimations is the requirement for measurements of current and future demographic base by location and estimates of current and expected changes in wealth. Wealth measurements are important in that they indicate the purchasing capacity of households for residential property, demand by households for goods and services obtained from commercial real estate structures, and real estate needs of business enterprises such as office space.

Real estate market analysis is at its best when assembling components required for the analysis and then producing analysis with those components that meet the objectives of the decision maker and market analyst. Just as estimation of real estate supply by product category has been revolutionized by the geographic technology industry, so too has estimation of demand. Real estate market analysts today are seldom required to derive primary data, such as measures of local employment or project local population trends. These measures are readily available from private data vendors.

Because private data vendors specialize in their activities and market niche, they have an advantage over the analyst in the construction of data sets. Private data vendors can allocate more resources in the construction of more accurate and more comprehensive data. Private data vendors have economies of scale from selling the data product to many end users. The comparative advantage that the analyst has is knowing the needs of the client and expertise in how to use data to meet those needs.

In this section of the chapter, several commercial data products are discussed that can be used in analyzing real estate demand. There are many such products on the market, and this is only a small sample.[13] These products are discussed as representative of niches within the geography technology market that provides commercial data products. The commercial data sets commented on here are from the categories of economic base and employment growth, demand segmentation using LSP and geocoding, and U.S. Census demographic and housing data.

Economic Base and Employment Growth

As economic base and employment expand, then demand will increase for all types of real estate products. There is no reason for a real estate market analyst to develop expertise, or allocate resources, for the construction of economic base and employment growth data. A variety of private data vendors provide these data. Below is a commentary on one such data set from Woods & Poole Economics.

Woods & Poole Economics Inc. CEDDS Data

CEDDS is an acronym for Complete Economic and Demographic Data Source.[14] CEDDS provides demographic and economic projections and historical data for all counties and Metropolitan Statistical Areas (MSAs) in the United States. The data can be imported into most spreadsheet and GIS software programs. To visualize and work with the data in a mapping software program, the user needs to have appropriate county or MSA geographic boundary files to which the CEDDS attribute data may be joined.[15]

Woods & Poole's county projections are updated annually. They use county models that take into account local conditions based on historical data from 1969 to the present. In the Woods & Poole model, the economies of counties are linked, so that the projected economic conditions in one county are reflected in the projected economic conditions in other counties.[16]

Demand Segmentation

Demand segmentation is the process of partitioning consumers into categories that allow for more accurate representation of those consumers' demand for real estate. As discussed in chapter 2, a variety of geodemographic measurements are available for this evaluation, including age and income. Lifestyle segmentation profiling is even more valuable because, in the construction of LSPs, it is recognized that people are complex combinations of many descriptive attributes, including stage of life cycle and income. An LSP combines many descriptive attributes into a single measurement. That measurement is used as the basis for demand segmentation. One private vendor for LSPs is CACI Marketing Systems. Among CACI's products in this market niche is their CACI Coder/Plus.

CACI Coder/Plus for Geocoding and LSP Segmentation

CACI Coder/Plus is a combined household lifestyle segmentation and geographic location software program that allows for a comprehensive assignment of geographic information (latitude-longitude coordinates and full FIPS (Federal Information Processing Standards) codes down to census blocks, and lifestyle segmentation indices based upon the best derived geographic coordinates.[17] Data fields with geographic and lifestyle segmentation information are added to the input data file.

CACI Marketing Systems provides demographic data and ACORN LSPs for the general business community. ACORN is CACI's acronym for A Classification of Residential Neighborhoods.

LSPs date from the work of geographer Brian J. L. Berry (1964). Berry was the first to propose the use of factorial methods for transforming social area analysis in his now-classic articles "Cities as Systems" (Berry 1964) and "The Factorial Ecology of Calcutta" (Berry 1969; see also Berry 1970). As a consultant, Berry was part of the development team employed by CACI to create the first commercial LSP.

The geocodes are used to look up the appropriate LSP for that location in the database. LSPs in CACI's database are assigned at the ZIP+4 code level, so adjoining

ZIP+4 codes might have different LSPs assigned to them, but all addresses within the same ZIP+4 will have the same LSP. Table 4.2 lists the 9 major LSP groups in the ACORN database and 43 subcategories. ACORNs and other competitive LSPs (Thrall et al. 2001) have become invaluable in understanding demand segmentation and customer profiles, whether they be Internet customers or traditional customers.

U.S. Census Demographic and Housing Data

CensusCD+Maps is a combination of demographic data and geographic software. Data include the complete U.S. Census of Population and Housing and TIGER/Line boundary files, down to the census block group level. The product does not include visualizeable streets. Data also include estimates and projections of demographics and consumer spending. Historical records include county statistics from 1790 to 1996. Software includes automated ring study report generation and visualization of the spatial data via a map generator and viewer. To execute a ring study, the program user enters into the software program either latitude-longitude coordinates or an address, and the inner and outer radius of the ring.[18]

Step 4: Compile Report and Present Analysis to Client

The final report should begin with a description of the overall project within the wide context of the economic, geographic, political, and social forces that are shaping the urban build environment (see Wofford and Thrall, 1997). The final report should then include a presentation of the findings on market area, supply of competitive product, and demand estimates for the product. The analyst then integrates these findings into his or her estimate of absorption, namely the amount of product over a specified period of time that can be built or acquired and then leased or sold (see Wang, Webb and Cannon 1990; and Heimsath 1991). The report should also include a statement of the goals and objectives of the client as those are understood by the analyst. The goals and objectives of the client will differ depending on whether the client is an investor, a developer, or a redevelopment agency.

The market analysis report generally concludes with recommendations on whether to proceed with the project. If there are cautionary notes, those notes should be included here as well. If the recommendation of the analyst is conditional upon a set of circumstances occurring, then those conditional notes must be included as well.

Conclusion

This chapter established the general framework for conducting market analysis for all types of real estate. With appropriate modifications, this general framework applies to all product categories of real estate. All real estate market analysis follows four steps: First, establish a trade area, also known as a market area, from which to draw the data

Table 4.2. ACORN lifestyle segmentation indices

ACORN Category	Description
1 Affluent families	1A Top one percent
	1B Wealthy seaboard suburbs
	1C Upper income empty nesters
	1D Successful suburbanites
	1E Prosperous baby boomers
	1F Semirural lifestyle
2 Upscale households	2A Urban professional couples
	2B Baby boomers with children
	2C Thriving immigrants
	2D Pacific Heights
	2E Older settled married couples
3 Up & coming singles	3A High rise renters
	3B Enterprising young singles
4 Retirement styles	4A Retirement communities
	4B Active senior singles
	4C Prosperous older couples
	4D Wealthiest seniors
	4E Rural resort dwellers
	4F Senior sun seekers
5 Young mobile adults	5A Twenty somethings
	5B College campuses
	5C Military proximity
6 City dwellers	6A East Coast immigrants
	6B Working class families
	6C Newly formed households
	6D Southwestern families
	6E West Coast immigrants
	6F Low income: young and old
7 Factory and farm communities	7A Middle America
	7B Young frequent movers
	7C Rural industrial workers
	7D Prairie farmers
	7E Small town working families
	7F Rustbelt neighborhoods
	7G Heartland communities
8 Downtown residents	8A Young immigrant families
	8B Social security dependents
	8C Distressed neighborhoods
	8D Hard times
	8E Urban working families
9 Nonresidential neighborhoods	9A Business district neighborhoods
	9B Institutional populations
	9C Unpopulated areas

for the real estate market analysis. Second, evaluate the competitive position of the project by estimating competing supply. Third, measure demand. Demand estimation includes the assembly and use of projections of economic, geographic, and social indicators that together influence the demand for a specific real estate project. Fourth, compile the report and present the analysis to the client.

PART II

APPLICATIONS TO REAL ESTATE PRODUCT TYPES

5

Housing and Residential Communities

Housing occupies about 70 percent of the land area of a typical city. That land area is not randomly distributed, but instead follows regular spatial patterns; these patterns are sectorial and radial (see Hoyt 1939; chapter 2). These geographic patterns form housing submarkets. Specific demographic groups are attracted to housing in those submarkets. As there are many kinds of demographic characteristics of households, there are also many types of housing, and many housing submarkets. Housing submarkets include downtowns, middle-burbs, suburbs; high income; middle income, and low income; new development, mixed use, older development, and mixed new infill with older development; apartments, condominiums; townhouses, high rises, and single-family dwellings. The market analyst makes recommendations on which type of development will be most successful in which submarket and on which submarket would be appropriate for a particular type of development (see Sumichrast and Seldin 1977).

Few people today choose to live without the benefit of some type of housing. The choice and availability of what type of housing to live in depends on a complex interaction of many factors, including culture, the natural and built environment, technological scale of society, government, income, stage of life cycle, economics of building construction, and knowledge and imagination of those building the housing. This chapter presents a broad overview of housing market analysis. In the overview, the determinants to demand and supply of housing are presented (See also Harvey, 1992). There is a broad overview of forecasting procedures and methodologies, the methods for projecting absorption rate, housing demand, and competitive supply, and how sales prices and rental prices might be determined.

Overview of Housing

Cultural and Institutional Changes

In the last quarter of the nineteenth century, upper-middle-income urban households in the United States and Canada often lived in what are today commonly referred to as Victorian houses. These houses were designed for multigenerational living, including grandparents as the head of household, their children, and their grandchildren. Aunts, uncles, and cousins might have lived in the same dwelling. All the family subunits contributed to the finances of maintaining the house. This provided social security to the elder members of the household, and inexpensive yet high-quality living conditions for the other family members. In a Buchanian theory of clubs sense, as discussed in chapter 2, such housing accommodations were a very efficient club. The extended family club influenced housing architecture, and housing architecture molded and transformed the family unit.

The United States and Canada in the twentieth and twenty-first centuries have been nations of migrants, quickly moving to those regions offering economic advantage, be they the oil fields of Oklahoma and Alberta or the Silicon Valley of California. Multigenerational in-place stability became economically inefficient for all but a few households. In other words, the extended family has become a disadvantageous club, and therefore housing to accommodate extended families became obsolete for that use. Other housing architectures were adopted, being better suited for the individual and family unit of today.

At the end of World War II, new households formed at a rapid rate, making up for lost time of the war and the Great Depression that preceded it. Those households were also prolific, giving rise to the baby-boom generation discussed in chapter 2. The economically depressed regions of the U.S. South and parts of the U.S. and Canadian Midwest sent migrants to the high economic growth regions of California and the Great Lakes industrial belt. The single-family dwelling became the housing of choice in the 1950s for the middle-income white nuclear family: husband, wife, and 3.5 children.

In the United States, government in the post–World War II era reinforced the household's preference for the single-family dwelling. The U.S. federal government biased the economy toward the provision and consumption of single-family dwellings. It was at this time that the geography of the U.S. and Canadian city became different, as the Canadian federal government did not offer the same bundle of incentives to producers and consumers of for-sale housing.

Among the subsidies that began in the post–WW II era were the G.I. Bill, Veteran's Administration (VA), and Federal Housing Administration (FHA) mortgages, and cheap land made available by massive highway and interstate freeway projects. As federal income taxes increased, and as many states piggybacked onto the federal income tax with their own income tax, the mortgage interest deduction provided further incentives for households to purchase housing (see table 2.3).

The supply of housing during WW II had not increased because materials and labor had shifted to production of goods required for the war effort. That alone might have been sufficient to create a housing shortage during the WW II era and immediately thereafter. However, in the post–WW II era, federal government regulations served to bias housing consumption toward new housing in the newly created suburbs. Federal legislation effectively made much of the pre–WW II era housing unqualified for VA and FHA loans.

Men and women serving in World War II, and shortly afterward in the Korean War, were entitled to low-interest VA loans for purchasing a house. Because the federal government was making loans, to insure its investment the house had to conform to certain federal government-specified building codes. Many of these codes were not in effect prior to WW II, and therefore much of the housing constructed before WW II did not conform to the new codes. Some of the housing stock built before WW II thereby became obsolete. The purchaser of such a house would run the risk of not having a future buyer. The result was that maintenance-based housing cycles (see chapter 2) led to neighborhood decline in many older American urban submarkets.

President Eisenhower had a vision of great German-like autobahns connecting the major cities of the United States, leading to the creation of the U.S. Interstate Highway System. The Interstate Highway System became the largest public works initiative ever in the United States. In terms of federal fiscal policy, the interstate highway public works construction effort would address the threat of rising unemployment. The public investment in the interstate gave access to undeveloped private land at the urban periphery where the single-family dwellings would be built. Construction of the interstates and single-family houses would put veterans of WW II and the Korean War to work.

The bias created by the U.S. federal government brought about a new geography to the urban landscape. Households became spatially dispersed in the new, low-density suburbs. The new geography also served the needs of federal fiscal policy objectives. Suburban living reinforced households' needs to consume large durable items, such as one or more automobiles, a clothes washing machine, and so on. The manufacture of these items, their support industry such as gasoline stations and repair services, and the consequences of the local multiplier effect (see chapter 2) resulted in even more employment opportunities and even greater demand for housing. The economy moved to a stage of high mass consumption (Rostow 1960). For further discussion see Harvey (1985).

The housing start index became a popularly accepted measure of transformation to economic prosperity for the U.S. middle-income household. At the beginning of the twenty-first century, housing starts are still reported on nightly television news broadcasts as a measure of the success of the economy and well-being of the nation. An economy with a high rate of housing starts has been viewed as one where there were substantial local economic multiplier effects and opportunities: the construction of infrastructure, construction of housing, purchase of washing machines, purchase of the vehicles for transportation, automobile maintenance and insurance, purchase of wall-to-wall carpeting, rising consumption of gasoline, and so on.

Architecturally, post–WW II housing followed Frank Lloyd Wright's vision of the prairie house: straight lines, low roof, simple and cost-effective to construct. That cost-efficient architecture was adopted in the Levitt-like towns across the suburbs of the United States. But these were mainly enclaves of white households.

Lower income households benefited indirectly from out migration to the suburbs during the 1950s. African-Americans leaving the restrictive South for jobs in the industrial Midwest and Northeast found inexpensive housing in the inner cores of those cities. This filtering process changed the demographic composition of the inner cores of U.S. cities.

Generally, the housing start index is a measure of local economic prosperity, but it can also in some instances be a measure of urban decay. For discussion of housing indicators see Miller and Sklurz (1986). During the last quarter of the twentieth century, the Buffalo-Niagara Falls, New York, Metropolitan Statistical Area (MSA) lost an average of 3 percent of its population each year. As the city deteriorated, maintenance-based housing cycles began in the older neighborhoods, leading households to flee to the suburbs in search of housing perceived as a safe investment. Because there was effectively no positive net migration to the MSA, housing starts could be interpreted as a measure of the rate of deterioration of the in-place housing stock.

The late 1960s was a watershed of social change. The young generation of women expected equality with their male counterparts. In 70 years, the American middle- and upper-middle-income family had changed from the Victorian era extended family to the stereotypical nuclear family household of the 1950s, and then became a far more varied family at the start of the twenty-first century. Today a family is inclusive of single-parent households, unmarried couples with or without children, and persons living alone, perhaps with transitory partners. More than 55 percent of all marriages end in divorce, and perhaps because of a desire to avoid divorce, the rate of marriage per cohort group has also declined (see chapter 2 on cohort groups).

There are several messages contained in the above introductory narrative. First, the type of housing we have depends on culture, construction technology, and government incentives and directives. Second, as any of these forces change, then so, too, does housing change. And these forces are changing. The size and design of the Victorian house, once the norm for some social classes, is now obsolete. Is the changing household, and aging of the population, leading the suburban single-family dwelling to the same fate as the Victorian house? Present trends do support the single-family dwelling becoming less dominant in market share of new housing construction. At the start of the twenty-first century, the single-family dwelling makes up 65 percent of all new housing, down from the decade of the 1990s when it was 81 percent. As the household becomes more varied in its composition, so too do the type of dwellings to accommodate the various subgroups.

Today the extended family is connected via e-mail; they do not necessarily live in the same house. Likewise, the household unit is changing. As the culture and associated family structure changes, then the type of housing that culture requires also changes. A suburban couple with two children becomes divorced and sells their suburban house. One spouse may move back in with his or her parents, thereby creating a new form of extended family. The other spouse may buy a smaller condominium nearer to work or downtown. The family culture is changing, and housing will adapt to the changes in the society. If the market analyst does not consider these cultural changes, then opportunities will be missed, or worse, housing might be built for which there is a declining market that then becomes less saleable and priced lower than expected.

Narratives like that above are useful for placing a proposed real estate project into correct cultural and urban geographic context. As discussed in chapter 4, the analyst's

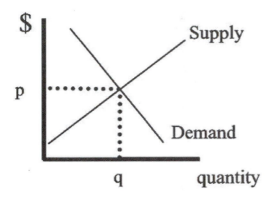

Figure 5.1 Housing supply and demand.

report must include appropriate descriptive narrative as subjective qualitative evaluation.

Some Economics

Heuristic housing models are based on observation and experience and generally conform to the logic of supply and demand curves. This section provides an introductory
narrative using the heuristics of supply and demand as shown in figure 5.1.

Housing Demand

Housing as a Normal Good   Housing demand is the amount of housing that households collectively within the market are willing to purchase at a given price. Normally,
the greater the price, the less housing will be purchased, and conversely, the lower
the price, the more housing will be purchased (see Thrall, 1983, 1987). When the
demand for housing behaves like the above as shown in figure 5.1, it is known as a
normal good.

As the number of households increase for any of the reasons discussed in chapter
2 as well as earlier in this chapter, then the demand for housing units increases from
$D[2]$ to $D[1]$ (figure 5.2). The price of housing then increases from $p[2]$ to $p[1]$. This
is the fundamental cause of the rise of housing prices in the United States from the
early 1970s through the mid 1980s. Zoning changes, infrastructure, and housing construction could not keep up with the increasing need for more housing units in many
markets, so the price increased, providing an incentive to developers to build more
housing.

Housing as a Veblen Good   There are instances when the demand as shown in figure
5.1 is not downward sloping. Thorstein Bunde Veblen (1857–1929), an American
economist and social commentator, wrote about the economic structure of U.S. society.
In his book, *The Theory of the Leisure Class* (1899), Veblen introduced the term
conspicuous consumption to describe the competition for social status. Veblin's conspicuous consumption has given rise to two specialized categories of demand: the

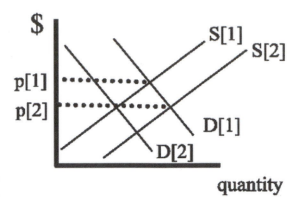

Figure 5.2 Identity problem in economics.

snob effect and the *bandwagon effect*. Veblen argued that certain goods confer status on the individual from their consumption, and one of the most important attributes of this class of goods, now known as *Veblen goods*, was price. Price itself could be a cause of demand. The higher price confers greater status from its consumption, and therefore the demand increases. In other words, as price goes up, then demand increases. Rolex watches, Mont Blanc pens, and Harley-Davidson big twin motorcycles might be Veblen goods. They are coveted not only because they keep better time, write better, or are mechanically engineered better than alternatives. Instead, these brands are coveted beyond the value received from the consumption of goods from their competitors because they also bring an added aura, mystique, and status, and therefore higher price to the good.

Housing also confers status on its residents by way of its architecture, and perhaps more important, its location. Sometimes, even having the right postal ZIP code will add value to a property by conferring status. Some purchasers might be willing to pay a premium for housing that confers the status desired and thereby bring about a market price affect. The same can be said about retail centers; a good purchased at an upper end lifestyle mall might bring a more pleasurable buying experience and therefore a high price (and higher rents for the retail center) than does the same good when purchased from a discount factory outlet center.

In some regions, a gated community is a characteristic of housing that might add a sense of exclusivity and might thereby create a measurable Veblen effect for housing in that neighborhood (see Barron 1998; Blakely and Snyder 1997). Residents might justify the premium paid for a house in the gated community on the basis that restricted access decreases the crime rate. However, the statistical evidence has yet to support that gated communities are safer than similar nongated neighborhoods, at least in the United States. So no real increase in security is added, yet even when presented with statistical evidence, some households are still willing to pay a premium to live in a gated community. The gates are a surrogate measure for something Veblenesque and difficult to directly measure. In residential market analysis, surrogate variables for perceived status, such as the presence of gates, are essential factors in the analysis.

Taste Preferences for Housing The residential market analyst identifies the set of characteristics on which households will base their decisions and the value those households place on those characteristics: the value of a gated community, the value of open-space and parkland, the value of a golf course. How the household values these characteristics depends on the demographics of the household: the stage of life cycle, age, health, income, and so on. For example, the more above the median income a household is, generally the greater is the value placed on quality of life and quality of environment. Conversely, the lower the income level, the less is the weight placed on quality of environment.

Lifestyle subdivisions conforming to sustainable urban development guidelines might not be an appealing way to allocate financial resources by lower income households. But a new development marketed to upper-middle-income households might be required to have the amenity and design characteristics of lifestyle subdivisions and sustainable communities; without those design characteristics, those households will not be interested in the development.

Taste preferences change. We characterize a decade by large style and preference trends of the time. These bandwagon-like style trends might give rise to "demand bubbles" of short duration for a style of housing, or even for a neighborhood. A housing characteristic that is a Veblen good, such as soaring eighteen-foot ceilings, might later become out of style if the cost of heating and cooling that empty space substantially increases. Of more lasting duration are the regional variation in taste preferences.

Homebuyers by region are seldom indifferent between a northeastern Cape Cod and a southwestern adobe. Regionally, architecture of each type has its supporters. The propensity for young households to own, versus rent, might be higher in one region than another, even though their demographic characteristics are the same. The same is true of retail: the food served at a Hawaiian poi restaurant might not be a big hit in Chattanooga. This argument can be extended to preferences, and willingness to pay, for amenities between regions, and even submarkets within a region. Residents of Longmont, Colorado take pride in not being like and not having a city like nearby Boulder. Therefore, the analyst must understand regional preferences and even preferences between submarkets of the same region.

The analyst must identify those demographic characteristics of the target population that the real estate development is to be marketed to. The analyst must also know how those demographic characteristics, which are actually surrogates for taste preferences and ability to spend, affect demand for various real estate products.

To accomplish this, the analyst creates a hedonic model of housing demand, following the same procedure as Applebaum's analog method presented in chapter 4. However, as discussed above, the analyst must add qualitative judgment to the report because such models are based on past market trends, and taste preferences can change, making past market trends unreliable indicators of future market estimates.

Economic Analysis of Demand and Supply Change What causes price trends? Change in demand? Change in supply? A problem with observing price trends and then attempting to explain those trends on the basis of movement in supply and de-

Box 5.1. Interpretation of housing price differences by demographic subgroups. An application of the identity problem

To complete a market analysis report, as in chapter 4, the residential real estate market analyst may draw upon the logic of figures 5.1 and 5.2. This example of the identity problem arises in the context of public housing issues and public policy analysis. Say a public housing authority or economic development organization requests comment on perceived housing market price differentials between two submarkets. The economists' identity problem may be a key for elucidating several alternative explanations. In scenario 1, the demographic subgroup is thought to pay a higher price for housing. In scenario 2, the demographic subgroup is thought to pay a lower price for housing.

Scenario 1: Explain why a demographic subgroup pays more for housing

First, on the supply side, if prices are indeed higher, how could this come about because of movement in the supply schedule for housing? Two hypotheses for the reduction in supply are that (1) real estate agents refuse to show housing to certain population subgroups except in specific neighborhoods, thereby lowering the supply of housing available to the subgroup, and (2) mortgage lenders are not willing to make loans to certain population subgroups except in particular neighborhoods, thereby making the supply of housing outside of those neighborhoods unavailable to the subgroup. In response to these hypotheses, federal legislation has created equal housing laws and the Community Reinvestment Act.

Second, on the demand side, the identity problem gives rise to an alternative hypothesis in support of a scenario of higher prices. Higher prices may persist by submarket because of a strong preference by a demographic subgroup to locate there. The resulting agglomeration might sustain business activities that otherwise might not exist, such as restaurants, religious facilities, and so on. In other words, the demand might be higher to reside in a specific submarket, thereby raising price.

Scenario 2: Explain why a demographic subgroup pays less for housing

Prejudice can be interpreted as one demographic group viewing another demographic group as a negative externality. The demographic groups might be aligned according to race, age, income, and so on. This logic dictates that demand declines as proximity increases to the group that is prejudiced against, thereby lowering the market demand for housing in that submarket. The price of housing declines within and near that submarket. Paradoxically, the group that is the target of the prejudice may possibly pay less for housing, but perhaps with the cost of a destabilized neighborhood. A surplus of housing in a given submarket can lead to a decline in prices, which can initiate a maintenance-based housing cycle.

mand is that movement in either supply or demand schedules can cause a change in price (see figure 5.2).

The market analyst may observe changes in market price for a type of housing product. Say the initial price observed is $p[1]$ and that price is observed to decline to $p[2]$. Was the decline in market price the consequence of an increase in the supply schedule to $S[2]$, while the demand schedule remained constant at $D[1]$? Or was the decline in price attributable to a decrease in demand from $D[1]$ to $D[2]$, without a change at all in the supply schedule? Observing price trends alone is not sufficient to establish the cause of the price trend. This is known as the *identity problem* in economics. Box 5.1 provides an example of how, because of the identity problem, alternative hypotheses about the cause of changing market conditions might be included in the report to the client,

Some Geography

How does a submarket develop through time? Geographers have developed the general reasoning known as *spatial diffusion* to explain how a phenomenon spreads across space through time. Descriptive discussion of spatially diffused phenomena dates to the latter part of the nineteenth century. Major advances in mathematically modeling spatial diffusion came about in the post–WW II era, primarily in the work of Swedish geographer Torsten Hagerstrand (1952, 1965, 1967a, 1967b). Hagerstrand believed diffusion to be a process in which anything that can change location originating from single or multiple sources, and in time will be spread through a space.[1]

The phenomenon initially diffuses at an increasing rate, reaches an inflection point, and then meets increasing resistance and thereafter increases at a decreasing rate, with further diffusion slowing down to a crawl, approaching an asymptote. The geometric form of this diffusion is known as a logistic S curve, as shown in figure 5.3. The asymptote in figure 5.3 is shown to be 100%, but might be a value less than 100%. The general equation for the logistic S curve is:

$$P = k/(1 + e^{a+bt}),$$ (5.1)

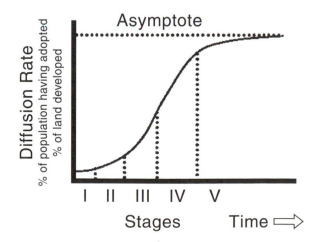

Figure 5.3 Submarket growth rate from diffusion processes.

where P is the rate of diffusion of the phenomenon, t is time, e is the base of natural logarithms, and k is the asymptote. The parameters a and b can be estimated using statistical procedures from the family of regression analyses; a is interpreted as the intercept and b as the slope of the line.

Diffusion is relevant here because, as outlined above, cities grow across space and through time. Geographic reasoning is that neighborhoods spread analogous to successive waves of logistic curves (Morrill 1965, 1968, 1970; Morrill et al. 1988; Thrall et al. 1993): First, there is the initial contact for development in a nearby submarket. There is resistance to the proposed development. Successive contacts break down that resistance. New development occurs in the nearby submarket, and the market penetration of the submarket increases until the submarket is saturated and thereby built-out. As saturation is approached, the diffusion process for new development begins in the adjacent submarket. Submarkets therefore grow in development like the logistic S curve in figure 5.3.[2]

Thrall and colleagues (1993) presented a model of urban development as a spatial diffusion process. Their cascade GIS (geographic information system) diffusion model was demonstrated for submarkets as small as a census tract for projecting the rate of absorption to have a typical error no greater than 3 percent. *Housing absorption* is the amount of housing that can be built and sold within a reasonable period of time. The following reasons summarize why the geographic approach (see figure 5.3) is appropriate for measuring housing absorption over a time period extending from first development to build-out:

- Housing absorption models generally use straight-line econometric forecasting procedures. These procedures project what has gone on in the past as continuing far into the future. Straight-line projection procedures do not match the actual rate of absorption, which is more like the logistic S curve in figure 5.3. Straight-line procedures do not take into account that submarkets become saturated with development and built-out.
- Urban development is like successive waves. As one submarket develops upward along the logistic S curve and more and resistance for further development occurs because of declining availability of land and increasing congestion, development spills over into adjacent submarkets, thereby beginning again the logistic S development process in that adjacent submarket.
- GIS technology allows for all property records for the entire market area and its submarkets for use in model calibration, allowing for unprecedented level of detail and accuracy.[3]

The following section is an outline of how one might conduct market analysis for housing. It draws together, as the market analyst must do, all of the appropriate material presented in the preceding background section of this chapter, as well as the material presented in the preceding chapters.

Conducting Housing Market Analysis

This section discusses how the general framework for conducting market analysis, as presented in chapter 4, applies to housing. This section references some of the technology and related issues of data. In chapter 4, the four steps for conducting all market analysis were presented: establish a trade area, determine competing supply, estimate demand, and compile a report.

Step 1: Establishing a Trade Area for Housing

Housing trade areas are most commonly established by using radial distance or drive time from a major employment center, or other important location. These rules of thumb are often used in combination with the concept of sectors presented in chapter 2. For example, if the builder desires to construct high-income housing, the market analyst might begin to define the submarket of interest by first identifying the high-income sector(s) and then honing in on smaller submarkets within those sectors. Similar procedures will be followed for moderate and low-income housing.

The large urban market can be decomposed along lines that follow geographic divisions established by government, the most common being postal ZIP codes and standard census divisions such as census tracts. The necessary data can be purchased for these geographic divisions. Local knowledge should be used to confirm that appropriate methods have been used to define the submarket.

Gainesville, Florida, and surrounding Alachua County are used for illustration. Local knowledge confirms that the high-income housing sector extends toward the northwest from the major employment centers. Say the developer owns land in that sector, in ZIP code 32605 (shaded in figure 5.4) However, ZIP code 32605 is a irregularly shaped polygon, so to define a submarket as that ZIP code would ignore important adjoining submarkets. Therefore, using the rule-of-thumb for radial market areas, a two-mile circle was drawn, which envelopes irregularly shaped ZIP code 32605, while also overlapping portions of adjacent ZIP codes. In practice, local knowledge might dictate larger or smaller radii to be used.

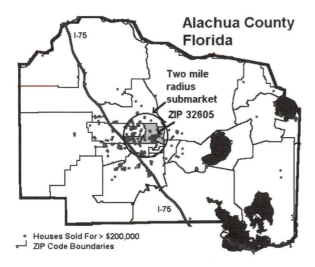

Figure 5.4 GIS view of prospective housing submarket.

Step 2: Estimating Competitive Supply

As discussed in chapter 4, for some real estate products, existing supply and supply in the pipeline can be acquired from F. W. Dodge and Comps.com and other data vendors (Thrall and Thrall 1991; Thrall 2001b). Supply data for apartments and mixed-use are available within those products, but not for residential housing. The best data for the current supply of housing and for deriving housing growth trends by submarkets like that discussed in figure 5.3 is property assessment data used for legal records and for calculating property taxes.

Most, but not all, U.S. states have open records laws, meaning that governmental records including property assessment data are available to the public. Property assessment data are normally collected and stored in compliance with recommendations set forth by the International Association of Assessing Officers. Those data records include transaction information on each property, such as dates of past three sales and their sale prices, owner's name and address, and descriptive characteristics of the property such as square footage, lot size, zoning, and parcel identification number. The address of the property is not always included, but the address where the tax bill is sent is always included.

A property assessment database will include all real estate within a county. Therefore, a data field is normally included that indicates the category of real estate product. For example, Florida mandates that all property records be coded with 1 of 100 possible Department of Revenue Codes (see table 1.1). If a county's property assessment records are included in a countywide GIS, then the data file will also include latitude-longitude coordinates or a parcel polygon geographic data file. For market analysis purposes, one latitude-longitude coordinate pair per parcel is generally sufficient.

In open-records states, property data can be obtained directly from the county property assessor's office. However, the data file format is often complex and not easy to work with. Data value-added resellers (DVARs) resell the data after it has been translated into a format that can be used on a desktop computer. Most DVARs also include a software program that can perform logical queries on the data to derive subsets of records to be used in further analysis.

There is no DVAR that provides property assessment data for all markets. Some DVARs are national but restrict the data coverage that they sell to the top-tier markets. Others are local, providing data only for a single market, and below the top tier. In some markets the analyst might be required to purchase the property assessment data directly from the local government, write software to translate the data into a format that can be used on a desktop computer, and write software to access the resulting data file.

Figure 5.5 shows a software application written in Visual Basic. The software allows structured query language-like queries of property assessment files and saving those records to a file that conform to the query. The software accesses data for the Alachua County example.

The software program user (figure 5.5) selects QBE (query by example), then enters into the display fields example values to form the query. To query houses, the user would input into the DORCODE field the value 100 because that is the Department of Revenue Code for single-family dwellings in Florida. To select houses that have sold for more than $200,000, the user enters >200000 in the field SALEVALUE. The

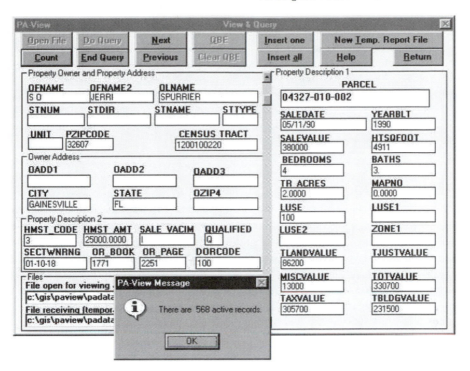

Figure 5.5 Example software program for querying property assessment data.

selected records, representing high-end housing for the example market, are exported to a dBase file for later importation into a GIS software program where the records can be mapped. This query returned 568 records that satisfy the 2 conditions. The market analyst must therefore either have knowledge of database management software, software programming, or have access to persons with that requisite knowledge.

Determination of housing in the pipeline is generally best obtained directly from building permit records. Within some markets, local DVARs collect and sell permit data. Housing and Urban Development publishes housing start data, but the delay time between collection and distribution decreases the value of this information. Also, to be comparable with the housing stock data, the housing pipeline data also need to be reported for standardized submarkets like ZIP codes or geocoded with latitude-longitude coordinates so that the analyst can place this information into the appropriate submarket.

Figure 5.4 shows a map of the records that were selected from the query of figure 5.5. Each record saved from the query is displayed as a point, overlaying ZIP code boundaries. The GIS software allows for point-in-polygon operations, so that if desired the number of points in each ZIP code can be counted, meaning in this instance that the count of the number of houses that sold for more than $200,000 within each ZIP code can be measured. Also, the square footage and median value by YEAR BUILT can be derived by ZIP code because these fields are available in the data set and therefore available for GIS logical and spatial operations.

Table 5.1. Characteristics by year sold

Sale year[a]	Avg. cost/SF	Avg. SF	Number
1	$84	3505	36
2	$93	3364	32
3	$83	3427	23
4	$86	3296	17
5	$77	4032	9

SF, square foot.
[a]Year 1 is the most recent year observed.

The GIS software of figure 5.4 was used to select 183 records from the subset of 568 records, which were located within the 2-mile radial submarket. Therefore, about one-third of all single-family dwellings that sold for more than $200,000 in the entire county are located within the chosen submarket. Summary characteristics of the 183 records are shown in tables 5.1 and 5.2. Table 5.1 shows characteristics of high-end houses by year sold within the submarket. Table 5.2 shows the actual absorption rate by year and the characteristics of those high-end houses built and sold.

Table 5.2 reveals that the target price point for new housing in the submarket is about $100 per square foot. The absorption rate for new high-end houses in the submarket has been nine per year, with a jump to eighteen in the most recent year. Because this is the observed absorption rate, then evidence is that the submarket has taken off for high-end houses in the example submarket. If demographic projections support further absorption rates at the level observed in the most recent year, then the increase in high-end housing within the submarket would be expected to grow like the logistic S curve as shown in figure 5.3.[4]

Step 3: Estimating Housing Demand

Business geographers performing market analysis are seldom required to derive primary data to estimate housing demand. Estimates are readily available from many private data vendors (Thrall 2001b). Table 5.3 is one such geodemographic report, generated with a combination of a software program and national database.

The commercial software product CensusCD+Maps was used to produce the geo-

Table 5.2. Characteristics by year built

Build year[a]	Avg. cost/SF	Avg. SF	Number
1	$97	3141	18
2	$96	3008	9
3	$100	3070	9
4	$89	3061	9
5	$93	3046	9

SF, square foot.
[a]Year 1 is the most recent year observed.

Table 5.3. Population and housing estimates: 29.6700 N, 82.3800 W; radius 0.00–2.00 miles

	1990	1997	2002
Population	28,706	31,639	33,645
Per square mile	2,087	2,300	2,446
Urban	28,706	31,639	33,645
Rural	0	0	0
Males	14,294	15,700	16,666
Females	14,412	15,939	16,979
White	26,005	28,615	30,481
Black	917	1,011	1,074
American Indian	28	34	33
Asian, Pacific	1,476	1,643	1,727
Other	280	336	330
Hispanic (any race)	1,647	1,828	1,970
Age 0–5	1,851	2,007	2,080
Age 6–11	1,880	2,133	2,293
Age 12–17	2,015	2,297	2,497
Age 18–24	5,166	5,448	5,664
Age 25–34	4,937	4,828	4,605
Age 35–44	4,489	5,193	5,656
Age 45–54	2,816	3,629	4,318
Age 55–64	2,214	2,394	2,573
Age 65–74	1,990	2,082	2,108
Age 75+	1,348	1,628	1,851
Families	7,244	7,968	8,377
Households	11,178	12,441	13,174
1 Persons	2,637	2,992	3,163
2 Persons	4,019	4,750	5,231
3–5 Persons	4,190	4,462	4,506
6+ Persons	238	237	274
Housing units	11,626	13,043	13,812
Occupied	11,084	12,439	13,173
Vacant	542	604	639
Owner Occupied	6,848	7,925	8,615
Renter Occupied	4,236	4,514	4,558

Consumer expenditures snapshot

	1990	1997	2002
Population		31,639	33,645
Families		7,968	8,377
Households		12,441	13,174
Expenditures total ($1000)		398,574	422,053
Housing, total ($1000)		140,975	149,275
Household furnishings ($1000)		19,092	20,215
Household operations ($1000)		5,392	5,710
Housekeeping supplies ($1000)		4,245	4,494
Shelter ($1000)		87,127	92,259
Utilities, fuels ($1000)		25,119	26,597

demographic ring study report of the submarket so defined. The report (table 5.3) shows changes in population and changes in housing expenditures within the desired radius.

Estimates for the two-mile circle that envelopes ZIP code 32605 shown in table 5.1 include an increase in housing expenditures by 5.9 percent during the five years following 1997, an increase of 690 owner-occupied housing units, and an increase of 733 households. In other words, the absorption rate of total housing units for the submarket is projected to be 690 units, meaning the number of units projected to be both built and sold. Some data products provide further breakdown by household income category, which would be important to include in a report on sustainability of absorption rates for high-end housing in the submarket.

Step 4: Compile Report and Present Analysis to Client

A market analysis requires a description of the overall project within the wide context of the economic, geographic, and social forces that are shaping the urban-built environment. The final report should then include a presentation of the market analysis findings on trade area, supply of competitive product, and demand estimates for the product. The report should place the project within the social, economic, geographic, and even political context of the community. The report should also include a statement of the goals and objectives of the client, as the analyst understands those goals.

Apartment Example

About 30% of households in the United States and Canada live in apartments (Bourassa and Grigsby 2000).[5] Apartments are not randomly distributed within a city, but are clustered into geographic submarkets (Hoyt 1939). Specific demographic groups choose to reside in apartments versus living in other types of housing such as single-family dwellings.

Here a business geographic procedure is introduced for conducting real estate market analysis for a new apartment complex. An example is provided for a new apartment complex in Gainesville, Florida. The four steps for real estate market analyses will be followed here.

Contemporary apartment complexes are generally built with several hundred units. These large complexes achieve economies of scale and therefore reduced per-unit costs of construction, maintenance, marketing, and administration. These reduced costs are necessary so that rental rates can be competitive with other rental complexes, and for sale housing. For discussion on how apartment rental rates are established, see Pagllari and Webb (1996).

Total fixed-cost capital expenditures rise with the number of units in an apartment complex. A modest 600-unit apartment complex, built with a cost of $40,000 per unit, will have a capital expenditure of $24,000,000. A 10 percent vacancy rate will translate into $2.4 million dollars of asset upon which returns are not being made, and fixed costs are not being covered. The cutoff to develop a new project may be in the range of 3 percent to 6 percent projected vacancy.

The business geographer will evaluate absorption rate of a proposed apartment if built within a particular submarket. The *absorption rate* is measured as the number of apartment units that can be rented within a particular price range. The submarket is a geographic area that identifies the location of the relevant competition and where renters will be looking for their accommodation.

Apartments are targeted to a specific target population niche. Competing apartments can be identified in a variety of ways, including geocoding databases of apartment addresses, visualizing the resulting mapped data, and calculating the lifestyle segmentation profiles (LSPs) of the apartments themselves (see Thrall 1998) that are derived using addresses of the apartments. The LSPs can be used as surrogate measures of apartment type and revealed target demographic niche. LSP can give the analyst insight as to the type and amount of amenities to include in the apartment development.

This apartment example was created for a developer considering a project midway between the University of Florida and downtown Gainesville, near a large hospital owned by the University of Florida. The university and downtown are about one mile apart. The developer specified that the architecture of the proposed 600-bedroom apartment complex would accommodate individual room-leases, with each apartment pod composed of four bedrooms and four bathrooms. Each apartment would include a living room and a kitchen. Management would lease each bedroom separately. The individual room-lease concept was chosen by the developer because projects of that type were thought to yield a higher rent per square foot than complexes where each apartment unit had a separate lease. The developer considered the target demographic niche to be all students enrolled at the University of Florida.

Four steps for conducting real estate market analysis are followed below: calculation of trade area, calculation of competitive supply, calculation of demand and target demographic niche, and presentation of recommendations. Each of these steps requires the use of GIS. Local sources of data required for this real estate market analysis are listed in box 5.2.

Step 1: Establishing a Trade Area for Apartments

Housing trade areas are most commonly established by using a rule of thumb such as calculation of a radial distance or drive time from one or several major destinations, such as employment and shopping centers. The sectorial nature of most cities must also be considered (Hoyt 1939). For example, if the target demographic population for a housing product is high income, then the trade area for the product would likely be in the sector known for that population group. The apartment market analyst might begin by first identifying the sector(s) where the target demographic group resides and then hone in on smaller submarkets within those sectors. See Bible and Hsieh (1996) for their analysis of apartments with GIS, and their regional market identification. Similar procedures are followed for moderate and low-income housing.

Since the target population was known to be students enrolled at the University of Florida, determination of the primary trade area was made on the basis of where these students live. All 45,398 students enrolled at University of Florida were geocoded using CACI Coder/Plus (Thrall 1998). Students were mapped as a point data file. Using ESRI's ArcView and their Spatial Analyst add-in, the kernel method was applied to visualize student density (Thrall and McMullin 2000). An envelope was drawn to inscribe the predominant locations of those students. The resulting trade area in-

---

Box 5.2.   Origin of data for apartment study

Student data
- Student address database from the registrar of the University of Florida
- Student address database geocoded using CACI Coder/plus
- Enrollment projections, year in university, gender, and student housing from the registrar and Office of the President, University of Florida

Apartment complexes
- Property assessment records of all real property in Alachua County provided by Ed Crapo, Alachua County property appraiser
- Property boundary geographic data file for all real property in Alachua County provided by Ed Crapo
- Apartments in the pipeline provided by City of Gainesville and Alachua County offices of "First Step" and building permits
- Addresses and identification of individual room lease apartments obtained from various guides to apartment living in Gainesville and surrounding Alachua County

For availability of apartment data, see Bogdon et al. (1999), and Thrall (2001b).

---

scribed more than 50 percent of all students enrolled at the University, and about 80 percent of students whose resident addresses were geocoded. The resulting primary trade area for students enrolled at the University is presented in figure 5.6. The high concentration, lower left inside trade area is an artifact of a large post office where many students choose to maintain post office boxes.

Step 2: Calculating Competitive Supply

Because of the uniqueness of the individual room-lease concept, the supply of similar apartment units with similar lease plans within the primary trade area was calculated. The competitive supply of other individual room lease apartments was derived from

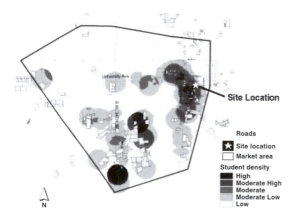

Figure 5.6 Trade area derived using kernel method for proposed 600-unit apartment complex, Gainesville, Florida.

Table 5.4. Competitive supply of individual room-lease apartments in trade area

Apartment complex	Address	Four-bedroom rate	Bath-rooms per apt.	Sq. feet	No. of bedrooms	Age (years)	WD furnished?[a]	Utilities
Campus Club	4000 SW 37th Blvd.	$440[b]	3	1400	922	4	Yes	Included
Campus Lodge	2800 SW Williston Rd.	$479[b]	4	1695	1116	New	Yes	Included
The Courtyards	1231 SW 3rd Ave.	$380	1.5	1200	374	28	No	Included
Lexington Crossing	3700 SW 27th St.	$445	4	1440	1020	2	Yes	Included
Melrose	1000 SW 62nd Blvd.	$412	4	1320	990	5	Yes	Included
Royal Village	710 SW Depot Ave.	$385	2	1088	456	3	Yes	Partial
University Club	2900 SW 23 Terrace	$340	4	1400	376	New	Yes	Optional
University Commons	2601 SW Archer Rd.	$315	2	1198	700	1	Yes	None
University Place	3705 SW 27th St.	$420[c]	3	1199	330	2	Yes	Included
University Terrace	3800 SW 20th Ave.	$350	4	1050	72	12	Yes	Included
University Terrace	3921 SW 34th St.	$350	4	1050	72	12	Yes	Included
Average rent per room		$377		1276.36	6428			

[a] Clothes washer and dryer provided in each apartment unit.
[b] Fully furnished; rate is not included in average rent per room.
[c] Three-bedroom unit.

listings in guides to apartment living in Alachua County. The tabulation is listed in table 5.4. The amenity package of each competitive apartment complex is listed in table 5.5. The proposed new apartment complex would need to be competitive in amenity offerings with apartments listed in tables 5.4 and 5.5. The proposed apartment complex would not include a pool, tennis courts, or computer lab.

The location of the existing competitive supply of all apartments, and individual room-lease apartments, within the primary trade area is displayed in figure 5.7. The map in figure 5.7 was constructed by importing the parcel boundary file for all prop-

Table 5.5. Amenities of competitive apartment complexes

Apartment complex	Pool	Tennis courts	Security system	Internet	Computer lab
Campus Club	Yes	No		Yes	
Campus Lodge	Yes	No			Yes
The Courtyards	Yes	No			Yes
Lexington Crossing	Yes	Yes	Yes		Yes
Melrose	Yes	Yes	Yes	Yes	Yes
Royal Village	Yes	No	Yes		
University Club	Yes	No	Yes	Optional	Yes
University Commons	Yes	Yes	Yes		
University Place	Yes	Yes	Yes		Yes
University Terrace	No	No	Yes	No	No
University Terrace	No	No	Yes	No	No

Blank fields = Unknown.

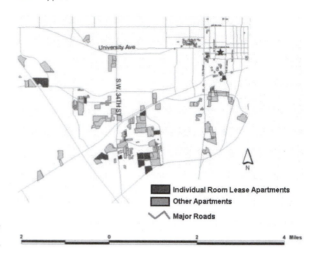

Figure 5.7 Competitive supply
of apartments in trade area.

erties within Alachua County. Each parcel included a parcel identification number (PIN). The Alachua County property assessment file was linked to the parcel data file by way of the common PIN field. The State of Florida Department of Revenue Codes for apartments were used to select those parcels that were apartments. It was thereby calculated that there is a competitive existing supply of 6478 individual lease rooms in the primary trade area.

City of Gainesville and Alachua County "First Step" and building permit records were reviewed to determine the supply of apartments in the pipeline (table 5.6). Pipeline supply is composed of projects that are not yet on the market but are at some stage of planning or development. The relevant supply is the supply that exists at the

Table 5.6. Apartments in the pipeline

Name	Address	No. of units	No. of bedrooms
Windsor Hall			88
Alligator Apt.	1012 SW 1st Ave.	6	
College Park	303 NW 17th St.	3	4
Cabana Grove	SW 20th/SW 62nd	624	1596
Plaza Royale	3802 Newberry Rd.	10	20
City Walk	715 NW 13th St.	120	474
Misc. Apt.	1806 W University Ave.		
Arlington Sq. P4	222 NW 1st Ave.	50	
High-rise multifamily	811 SW 11st		
Townhouse		6	
Multifamily	110 NW 7th		
Multifamily	1800 SE 8th Ave.		
Apt.	1103 SW 4th Ave.		
Apt., mixed use	503 SW 2nd Ave.	250	520
Hidden Lake Apt.	NW 21st		
Total		1069	2702

time the new development is completed and for a reasonable time horizon thereafter. Therefore, competitive projects in the pipeline must be considered. (For more on pipeline and commercial pipeline data, see Thrall 2000.) Table 5.6 reveals that there are at least 2702 apartment bedrooms within the primary trade area in the pipeline. The data do not reveal which units will be individual room-lease. Also, because many of the projects are at early stages of planning, the exact number of apartment units and number of bedrooms are not known for each apartment complex. Table 5.6 is an estimation of the number of apartment bedrooms that will be built. Some of the apartments listed in table 5.6 will never make it to the final stage of certificate of occupancy.

Step 3: Calculating Demand

Off-campus, individual room-lease apartments are not equally attractive to all student demographic niches. Students' addresses are available in the student record data file. Addresses of competitive individual room-lease apartments are known. Students were then selected who resided at addresses of individual room-lease apartment complexes. The student records contained students' year in college and other information. Figure 5.8 is the resulting breakdown of students by year in college who reside in individual room-lease apartments.

Students at the University of Florida generally prefer to live on-campus during their first two years. Students at the freshman and sophomore levels generally live off campus only if campus housing is not available. The pie chart in figure 5.8 reveals that graduate students have a strong preference against individual room-lease apartments. It had been hypothesized that more women than men would reside in individual room-lease apartments; however, the data demonstrated that the same proportion of men and women resided in individual room-lease apartments as were enrolled at the university.

The proportion of University of Florida students residing in the primary trade area was calculated by year in university. As year in school increased, the proportion of students residing inside the primary trade area decreased (see table 5.7). Forty-six percent of graduate students are estimated to reside outside the primary trade area shown in figure 5.6.

The University Registrar provided the figures for table 5.8. According to the Office of the President of the university, the official policy for the five years following 2000

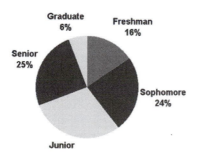

Figure 5.8 Characteristics of students in room-lease apartments.

Table 5.7. Students within and outside primary trade area (PTA)

	No. in market area	Percent residing outside PTA	Students with an Alachua Co. address
Freshmen	6010	9	6534
Sophomores	4132	15	4765
Juniors	5703	27	7251
Seniors	5247	32	6900
Graduates	4387	46	6419
Other	210	58	331
Total	25689	25	32200

is for significant increases in graduate student enrollment, while undergraduate enrollment will have very small increases, and perhaps even decreases. A decline in 229 students are projected for the lower division (i.e., the freshman and sophomore levels). The upper division (i.e., juniors and seniors) is projected to increase by only 452 students over the 5 year period. If past trends continue, 46 percent of the increase in graduate students would be expected to reside within the primary trade area.

Step 4: Summary Absorption and Recommendation to Client

Box 5.3 provides a summary of the report presented to the client. The analysis demonstrated that there is too much competitive product in the pipeline. The increase in competitive supply will be targeting a constant or declining population. Therefore, vacancy rates are expected to increase, followed by a possible decline in rental rates (Frew et al. 1990).

The recommendation is that the developer abandon the 600-unit individual room-lease apartment project. If the development proceeds, it is recommended that the amenity package be improved to be competitive with existing supply. The developer is also advised to change the target demographic niche to include graduate students. Graduate students were revealed to prefer other forms of housing than the individual room-lease design. The change in target demographic niche would require a change in the architectural design and a change in management of the completed apartment complex.

The value that GIS brought to the above real estate market analysis for a new

Table 5.8. Projected student increase by year in school for next five years

	2000	2001	2002	2003	2004	2005	5-Year change
Lower division	13253	13352	13024	13024	13024	13024	−229
Upper division	18549	18655	19139	19216	19001	19001	452
Subtotal	31802	32007	32163	32240	32025	32025	223
Graduates	8681	9131	9581	10031	10481	10931	2250
Total	40483	41138	41744	42271	42506	42956	2473

Box 5.3.   Summary of report to client

Positive aspects of apartment development
- Excellent access to university, hospital, downtown entertainment
- Good development potential because of existing high student density
- Good public transportation access

Negative aspects of proposed apartment development
- Apartments in the pipeline are moving into this market area
- 1000 bedrooms are estimated to be in the pipeline within three blocks of the proposed apartment development, not including the proposed 600 individual room-lease units
- Number of undergraduates will be increasing by only 223 students over the next five years
- Graduate students will be increasing by 25%; however, graduate students prefer other types of housing, and a large proportion of graduate students have traditionally resided outside primary trade area
- May result in cannibalization of existing supply of apartments

Suggested changes to development plan
- Reconsideration of individual room-lease to other administrative mechanism and architecture that would attract wider segment of the population, especially graduate students and entry-level professionals
- Consider mixed use—retail with apartments
- Reconsider a more comprehensive and competitive amenity package

---

apartment development includes increased accuracy, increased productivity allowing for increased timeliness, and overall improved analysis over pre–GIS methods; the judgmental decision should thereby be improved. See also Sriram and Anikeeff (1991) for discussion of judgmental issues in multi-family housing.

The recommendation in this instance is that the developer not proceed with the project as proposed. This GIS-based real estate market analysis might have saved the developer from making a $24,000,000 error in judgment.

Conclusion

This chapter presented background to the economics and geography of the housing market. The general framework for conducting market analysis for housing was applied to the four steps of market analysis. The four steps are establish a trade area, evaluate competing supply, measure demand, and present final report.

The examples in this chapter relied on a commercially available data product for the geodemographic analysis and GIS software in combination with a custom software application written in Visual Basic for the derivation of absorption trends and for defining the competitive position within the submarket. This underscores that the market analyst of the twenty-first century must have knowledge of the geographic technology industry as well as facility in the manipulation of databases, software, and even writing computer programs.

6

Office and Industrial

A developer needs advice on the market for commercial space, including office and industrial properties. An owner of a commercial building needs to determine how much to charge for leased space, how much to sell the property for, or how much the property can be refinanced for. A purchaser needs to determine if market conditions support purchasing commercial space, or renting, and at what price. The real estate market analyst is responsible for the creation and assembly of information to guide such decisions. A background overview of real estate market analysis for the product categories of office and industrial projects is presented.

Background

Hedonic Approach

The hedonic approach hypothesizes that a variety of phenomena contribute in one way or another to determining market rent.[1] In a hedonic model, office or industrial property rent or occupancy rate may be the dependent variable of a regression equation, as explained in chapter 4. The phenomena that are hypothesized to cause the value of the dependent variable are the independent variables of the regression equation. Some examples of independent variables that have been hypothesized and examined in hedonic models as to their contribution to determining office market rent are listed below:[2]

- Terms of lease (Glascock et al. 1990).
- Architectural design (Hough and Kratz 1983)

- Building characteristics (Vandell and Lane 1989)
- Access to white collar employment (Clapp 1980)
- Local property tax rates (Wheaton 1984)
- Status and prestige (Archer 1981; Archer et al. 1990)
- Agglomeration—benefits of high geographic concentrations of specialized office establishments for specific kinds of industry (Gad 1979; Kroll 1984)
- Spillovers from close geographic proximity (Clapp et al. 1992).

Hedonic models might also include dummy variables as independent variables to represent the presence of some characteristic or phenomenon. The dummy variables have an assigned the value of 1.0 to denote the occurrence of some characteristic and 0.0 to denote its absence.

Hedonic Models and GIS

An expectation must be developed by the analyst on how markets and submarkets differ in their rents, vacancy rates, and absorption rates and what their trend is expected to be. This intuition on office and industrial markets comes from a combination of experience and from the discussion of the spatial equilibrium heuristic models (see chapters 2 and 3 and Thrall 1991). Each submarket may be at a different period in its trajectory toward spatial equilibrium adjustments because of local, regional, or national market conditions. The analyst then anticipates the trajectory of change of rents, vacancy rates, and absorption that come about as the local market or submarket responds to market changes. Hedonic models are used to assign numbers to hypothesized trajectories of the market. (For additional hedonic price equations, see Glascock et al. 1990; McDonald 1993; Mills 1992; Wheaton and Torto 1994).[3]

The hedonic regression approach has been valuable in documenting the importance of a variety of phenomena, including geographic phenomena, and the value that each contributes to the real estate asset. However, hedonic models fall short because the results are merely correlational and do not provide causal evidence. To substantiate causal relationships, there must be a close integration of the heuristic general theory, such as that presented in chapter 3 and Thrall (1991), with the hedonic equations used for estimation. The real estate market analyst normally places this qualitative integration into step 4, the overview report to the client, and it is also included when specifying the hedonic equations for demand and supply in the general market analysis steps 2 and 3, respectively. Not only must the hedonic model be consistent with general theory, but to document the importance of the independent variables and to accurately depict the trajectories of the dependent variables, all the terms in the equation must be specified accurately and consistently with real world phenomena and behavior. Box 6.1 summarizes some of the problems associated with measuring what office rent actually is. Also, later in this chapter, problems associated with measuring geographic relationships are explained.

Hedonic models are not necessarily the same for each market and submarket. To learn how to design hedonic models for real estate market analysis, it is useful to look to the peer review literature written by scholars of the field. Five such analyses are summarized below, followed by a descriptive model of land use succession which is input into the qualitative evaluation included in the final report.

Box 6.1.   Lease terms

As space available increases, then lease terms become more important to rental rates. Instead of building management lowering rental rates, terms more favorable to the tenant may be adopted. For a discussion of the relationship between office rental rates, vacancy rates, and absorption, see Frew and Jud (1988), and Thompson and Tsolacos (1999). This strategy maintains a consistency in rental rates between the various tenants within a large office building and between office buildings located in the same submarket. However, by providing such non rent benefits to a new tenant, management of an office building is more likely to attract the new tenant to the available office space. The following are examples of items that, depending on how they are negotiated by tenant or building management, shift real rental rates upward or downward. An analyst that only looks at office rental rates may not be viewing the real rental value.

- **Prepayment:** Building management may request prepayments in a tight market or rent paid quarterly in a loose market.
- **Fit-out:** Building owners may pay for the cost of fitting-out the office space or credit the cost of fitting-out as reimbursement in rent.
- **Operating costs:** Operating costs may be included with base rent or passed straight through to tenants.
- **Minimum term:** Tenants may be offered short leases with options to extend at the current lock-in terms.
- **Security deposit:** A security deposit can range from no deposit to several months' rent.

Clapp's Hedonic Model of Office Markets

The first example is the hedonic model of John Clapp (1980). He created a database of 105 office buildings in Los Angeles. His database included three measurements of location: distance to the central business district (CBD), average commuting time of the building's workers, and square footage of office space within a two-block radius. The nonlocational measurements included in Clapp's hedonic model and database were descriptive characteristics of the 105 office buildings and annual rental rates per square foot of office space. Clapp found that distance to the CBD was the most important locational factor in the determination of office rental rates. Clapp's interpretation of this was that firms were willing to pay higher rents to locate within the CBD to have access to agglomeration economies.

Cannady and Kang's Hedonic Model of Office Markers

In a subsequent and similar study to that of Clapp. (198), Cannady and Kang (1984) created a database of nineteen office buildings in the Champaign-Urbana, Illinois, urban area. They found that straight-line distance to the largest employment and activity center in the market, in this case, the campus of the University of Illinois, was the most significant locational variable. Their locational variable, straight-line distance, overwhelmed the other locational variables in their hedonic model (see discussion of stepwise regression in chapter 4).

Cannady and Kang's work is important in that it established a point of reference to discuss how to measure locational variables. The real estate analyst that develops a hedonic model must consider the rationale for each variable. In the presentation of the analysis, the analyst must include discussion that justifies that the units of measurement of the variables are appropriate for the analysis. This is especially true for units of measurement of locational variables.

Cannady and Kang's use of straight-line distance measurements to measure proximity is today considered as possibly introducing large errors into the hedonic equation (Thrall 1998). Proximity to clients and business services is considered by many analysts to be the most important overall locational concern for office users. Proximity should be measured in the same manner as office users are making locational decisions. Usually, those decisions are made in terms of time required for access, or simply transportation time such as driving time or walking time. Straight-line distance measurements do not take into account barriers such as railroad tracks or rivers, nor do they take into account the many obstacles that must be traversed, such as a city block or a building. Seldom can one actually go in a straight line in a city. Therefore, a more relevant measurement of proximity within a large urban market is estimated transportation distance or transportation time.[4] Geographic information system. (GIS) technology allows for the estimation of travel distance and travel time between one or more origins and one or more destinations.

Sivitanidou's Hedonic Model of Office Markets

Rena Sivitanidou (1995; also see Sivitanidou and Sivitanides 1995) improved upon the hedonic model of Clapp and that of Cannady and Kang. Her more exhaustive hedonic model of the locational determinants of office rents uses a database of 1462 office buildings located in the Los Angeles area. Her hedonic model was based on the hypothesis that office rents depend on the three general concepts listed in table 6.1.

Table 6.1. Geographic measures of the determinants of office building rents

General concepts to be measured	Advantages the location brings to firm productivity	Worker amenities or disamenities	Local institutional controls
Geographic and attribute data used to measure general concept	1. Distance to the CBD 2. Distance to the nearest airport 3. Number of freeway miles passing through the commercial district where the building is located.	1. District's crime rate 2. Retail employment per resident population 3. Education expenditures per student by school district 4. Distance to the ocean	1. Commercial office zoning 2. Density constraints and presence of growth moratoria on office–commercial development

Compiled from Sivitanidou (1995).

Measurements used to represent each of these three concepts are listed in the rows under each column in table 6.1. When she calibrated her hedonic model, she found that the following variables were statistically significant: distance to the CBD, distance to a freeway, distance to the closest airport, education expenditure, crime rate, distance to the ocean, and retail employment.

Howland and Wessel's Hedonic Model

Howland and Wessel (1994) presented a database management procedure, including GIS, to project growth of office demand. Prince George's County, Maryland, was chosen to test their methodology. There are three steps to their database process-driven hedonic model.

They first assigned latitude-longitude coordinates to office buildings. The address of the office building was used to derive the geographic coordinates. The analysts also had available data on the inventory of office space within each building.[5] Next, a database of business establishments was created and geographic coordinates were assigned to the business establishments as well.[6] Included with the business establishment data was number of employees by business establishment. Business establishments were categorized by two-digit SIC (Standard Industrial Classification) code. Because they had a database of office inventory and number of employees by business establishment categorized by SIC code, they were able to estimate the proportion of employees located in office buildings at the two-digit SIC code level.

Their final step was to calculate office space demand based on derived space required per employee by two-digit SIC code. Employment growth rate was assumed to continue at the rate for the six-year period starting in 1980. To calculate projected office space demand, the growth rate was multiplied by space required per employee by two-digit SIC code.

Howland and Wessel's three-step method requires established office submarkets, including sufficient inventories of office space in those submarkets to create reliable data sets. Also, the method depends on reliable employment growth projections. Significant error in projecting demand for office space occurred because of an unanticipated slowdown in employment growth in the county during the period that was being forecast.

Bollinger, Ihlanfeldt, and Bowes's Hedonic Model

Bollinger and colleagues (1997) hedonic model was based on the belief that the most important determinants to office rents is the relative location of the office building to other phenomena. *Relative location* is a geographic concept measuring distance between phenomena on the landscape in some relevant units of measurement, such as travel time or street mileage. This is in contrast to absolute spatial location, which is measured in geographic coordinates such as latitude and longitude.

Among the phenomena hypothesized to be important in the Bollinger et al. hedonic model that operationalized this concept of relative location are proximity to clusters of office buildings, also known as agglomeration; proximity to transportation systems such as rail and interstate; and proximity to the residential locations of office workers. Bollinger et al. were not the first to hypothesize that agglomeration (i.e., geographic clustering on the landscape) was important. Indeed, these concepts were known in the nineteenth century for both retailing and manufacturing.[7] Their model did, however, confirm using contemporary accepted procedures that relative location and agglomeration contribute to office rents.

Land Use Succession

The preceding discussion has dealt with the concept of *situation*, presented in chapter 2, as the geographic relationship between an office building and office market to some important point destination, such as an airport or CBD. But situation deals with more than points on a map; it also refers to the context of the development to the large geographic patterns of urban land uses as discussed in chapter 2. With few exceptions, office markets do not come about independent of other urban land uses.[8] There is a normal succession of urban land uses, including the development of submarkets of offices.

GIS has been used to demonstrate the succession of urban land uses—to visually depict a historical geography of the urban development of counties from their beginnings to the present (Thrall et al. 1995).[9] Today, the succession of land use is generally housing construction, retail, and then office development. Finally, the new retail and office employment act as a generator for further housing development nearby. For further discussion see Garreau (1991).

Heuristic Theories and Hedonic Modeling

Heuristic theories and hedonic modeling can be compared to an FM stereo receiver. There are a finite number of controls for each receiver. The FM stereo will work in Orlando and in Denver, the difference being where the knobs are set between the two markets. Likewise, the numbers of market forces are finite that affect any real estate market, including office and industrial. However, the strength and importance of any particular variable will be different in each market and submarket. Measuring the strength and importance of hedonic variables via the process of regression is referred to as *calibrating the model*.

Calibration will result in different regression coefficients in each market, analogous to setting the stations on an FM stereo at different frequencies. But unlike the FM stereo analogy, there are no regulators ensuring how market forces are manifested. Generally, the calibration will measure the differences in market forces. However, it is possible that particular submarkets have some unique characteristic that may not have been considered in the variable specification of a general hedonic model. Therefore, to confirm that the appropriate variables have been included in the hedonic model and to confirm that they have been specified appropriately (i.e., straight line vs. drive time), it is recommended that the analyst either confer with experts familiar with the local market or make a personal, onsite inspection of the market. An honored tradition within the discipline of geography is to refer to onsite inspection as "muddy boots" geography (Thrall, 1998).

Business geographers performing market analysis on commercial real estate should provide information concerning the space available, market demand for that space, rental rates, the terms under which leases are negotiated, and the quality and location of the buildings offering those rates, vacancy rate trends, construction pipeline, nature of the space, nature of the firms in the area, and so on (see Sarvis 1989).

Real estate market analysis, then, is a combination of hedonic modeling, heuristic modeling, and qualitative analysis of expert opinion. As in the previous chapters, the general four steps are followed below in calculating market position of a project.

Step 1: Establishing the Trade Area

An analyst would not presume that a particular cluster of office buildings is a good investment based on national forecasts, or even on countywide measurements that indicate the county has good growth potential because (1) the growth that takes place in the larger region may be outside the market area of the particular submarket cluster of commercial buildings, and (2) the particular cluster of commercial buildings might be a submarket undergoing a downturn counter cyclical to the growth of the larger region. These considerations highlight the importance of appropriate geographic scale and identifying the relevant trade area, as discussed in chapter 4.

Office buildings cluster around nodes of other economic activity (Harris and Ullman 1945). Cities generally have multiple nodes of commerce, with each commercial node often being dominated by a narrow band of activities such as hospitals, universities, retail, professional services, and so on. An office submarket can be defined and explained as a variation of the definition of a submarket in chapter 2. An office submarket is a geographic area where

- Office buildings are clustered, thereby providing agglomeration benefits, or even disadvantages, to other office buildings in the cluster or the submarket as a whole.
- Office buildings within the same submarket will mostly be at the same stage of cycle, but perhaps out of phase with other office submarkets. Thus one submarket could be declining while another in the same market area is on the ascent.
- Land use within the office submarket is homogeneous and differs from surrounding land uses. Office space dominates the submarket. However, the office submarket is generally highly interdependent with the surrounding land uses.
- A submarket of offices comprises buildings of similar vintage, size, architecture, or buildings that have been retrofitted so that any differences in these characteristics are unimportant. Therefore, for the large part, within an office submarket, office space in one building is substitutable for office space within another building of the same submarket.

Cities have one or more office submarkets. Each office submarket is affected by the whole city, including the economy of the city, the other office submarkets, and submarkets of other land uses. Office submarkets are highly interdependent with the geodemographic characteristics discussed in chapter 2. Because office space users are drawn from the larger market area of the city, the office submarkets may be highly interdependent with other office submarkets that are far away. This interdependency leads to prices and absorption rates between submarkets being linked; an oversupply of office space in one submarket can have a price effect on another distant office submarket.

The geographic boundaries of an office submarket are established with a goal of minimizing the variation in the descriptive characteristics and phenomena that char-

---

Box 6.2.   Class A and B Space

By convention, office space is categorized into classes denoted as A, B, and C, with A being the highest quality and C the lowest. The standards to designate an office building as a particular class differ from market to market. Class A office developments generally meet international standards of design, construction, and facilities management and have a prime location. A prime office location generally has access to other existing high-quality commercial and retail developments, government offices, prime residential neighborhoods, major private transportation arteries, and high-speed public transportation.

An oversupply of office space generally leads the market to set higher standards for office space defined as class A. Changing expectations by tenants and loose market conditions also lead to higher requirements for inclusion as a class A building. Changing requirements might include being wired for local area networks and high-speed Internet access. Loose market conditions bring about higher expectations for modern, user-friendly, aesthetically appealing design and well-executed construction with international-quality building materials and building systems. Class A buildings are also expected to have attentive property management, security and amenities such as parking, high-speed elevators, and dining, and in some markets even exercise facilities.

There is no exact definition of a class B office property. Generally, if a property lacks one or two features described above, it is relegated to the class B category. Class C property would be deficient in many of the above features.

A building constructed to be class A might, with changing market conditions, changing maintenance, or simply aging, become class B, and then class C.

---

acterize the office submarket. Therefore, two office submarkets can be near to one another and even adjacent to one another. However, for this to be justified, there would need to be some significant characteristic that distinguishes the two office submarkets. Distinguishing characteristics generally arise from historical-geographical reasons. One submarket might have developed in the nineteenth century as factories. Subsequently that submarket may have been renovated into office space, while retaining the historical character of the neighborhood. The adjacent and distinctly different submarket might be composed entirely of contemporary twenty-first-century office space. The final test for justifying the two adjacent neighborhoods as being distinct submarkets is whether the tenants from one market consider office space in the other neighborhood as a near-perfect substitute.

The larger the submarket, the more variation there will be within the submarket. The smaller the submarket, the less the expected variation among sites. On the one hand, office submarkets should not be defined so small as to have measurable substitution of office space by office users between two adjacent submarkets. On the other hand, office submarkets should not be defined so large as to inappropriately combine neighborhoods where no meaningful substitution could occur from one neighborhood to the other.

In practice, the geographic delineation of office submarkets generally follows a combination of visual observation and the rule-of-thumb method as presented in chapter 4. Because the land use within an office submarket differs so much from surrounding land uses, the land use itself can then be used to geographically delineate the office submarket. Where are the office buildings concentrated? Draw a boundary envelope around the office buildings. That is the office submarket. There are three steps to delineate office submarkets using geographic technology.

First, within a GIS software program, display a point or polygon map of office buildings. A variety of criteria might be used to select a subset of office buildings to display on a map, including office buildings restricted to being valued at more than $250,000, taller than two stories, or greater than 125,000 square feet. Some GIS software can even display the footprint and floor plan of an office building as a three-dimensional image that can be rotated to represent any viewing angle.[10] Second, within the GIS software, overlay the map of office buildings over a satellite image or air photograph of the market area. Third, the business geographer, using his or her judgment, draws a polygon to inscribe the appropriate office buildings, defining the boundary of the submarket and digitizing that boundary. This process might also be automated using algorithms like those for figure 4.1.

The above definition and derivation of an office submarket is not equivalent to a trade area for the office building. An office building or office submarket serves a larger trade area, just as retail facilities serve a trade area and submarkets within those trade areas. By analogy, the employees within an office building or office submarket can be mapped using Applebaum's customer spotting method discussed in chapter 4. Instead of customer density, the business geographer would map employee density using the kernel method (see chapter 4). The primary trade area of an office building or office submarket may be defined as the geographic area that contains 80 percent of the employees.

An example of Hillsborough County, Florida, is used to illustrate the four-step procedure for office market analysis (Thrall and Amos, 1999a, b, c, d). Figure 6.1

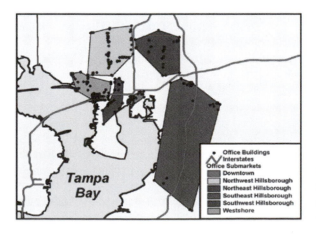

Figure 6.1 Submarkets of Hillsborough County (Tampa), Florida. From Thrall and Amos (1999a, b, c, d).

shows office submarkets of Tampa and surrounding Hillsborough County, Florida. The data for the supply of office buildings used to illustrate the concepts here, for the Hillsborough County example, were obtained from *Black's Guide to Office Space* (1997). Another source of office supply data, as discussed in chapter 4, is Comps.Com for existing inventory and F. W. Dodge for the pipeline. Now that office submarkets for Hillsborough County have been geographically defined (see figure 6.1), proceed to the next step of calculating the competitive supply. A demonstration of F. W. Dodge and also Comps.Com data and software are included in the supply analysis section. The supply of office space using data from *Black's Guide to Office Space* is also provided.

Step 2: Supply Analysis

Supply analysis must include an inventory of all relevant properties that are existing and occupied or available for rent, and of competitive properties in the pipeline. A descriptive inventory of existing facilities should first be created and broken down into the relevant geographic scale and submarkets.

Deriving Pipeline Data with F. W. Dodge

Deriving the supply of office, warehouse, or industrial space, at some stage of the development pipeline was at one time a very arduous task. It required the real estate market analyst to extract the necessary information from building permits. Today, for the larger U.S. markets, acquisition of pipeline data requires only a subscription to access the data via the Internet. The business geographer then can place the pipeline data into GIS software for subsequent submarket analysis using applicable submarket boundaries, and perform queries that allow him or her to zero in on supply in the pipeline that is competitive with a proposed project. The following is a demonstration.

First select the product type that would be considered as a competitive product for the proposed project. The F. W. Dodge options are shown in figure 6.2. Second, select an approximate trade area. This might be countywide, MSA (Metropolitan Statistical Area), or combination. Selection of a geographic region larger than the actual trade area is advised; after the data are imported into the GIS software, the competitive projects can be selected on the basis of being contained within the geographic boundaries of the actual trade area. This is shown in figure 6.3.

Third, competitive real estate projects can be further refined within the F. W. Dodge structured query language-like software, or may also be accomplished after the data are imported into the GIS software. Among the considerations for refining projects would be particular product brands or developers that are known to directly compete with a unique property type of the proposed project. This is shown in figure 6.4.

Fourth, the analyst can view the selected projects (figure 6.5) from the F. W. Dodge database and export all or a subset that have been selected to data formats that can be read by spreadsheet and GIS software. The most universal data formats for this purpose are Microsoft Excel (.xls) or dBase (.dbf).

Fifth, derive summary measurements of development activity of the chosen real estate product category in the primary trade area. If the area defined in figure 6.3 is

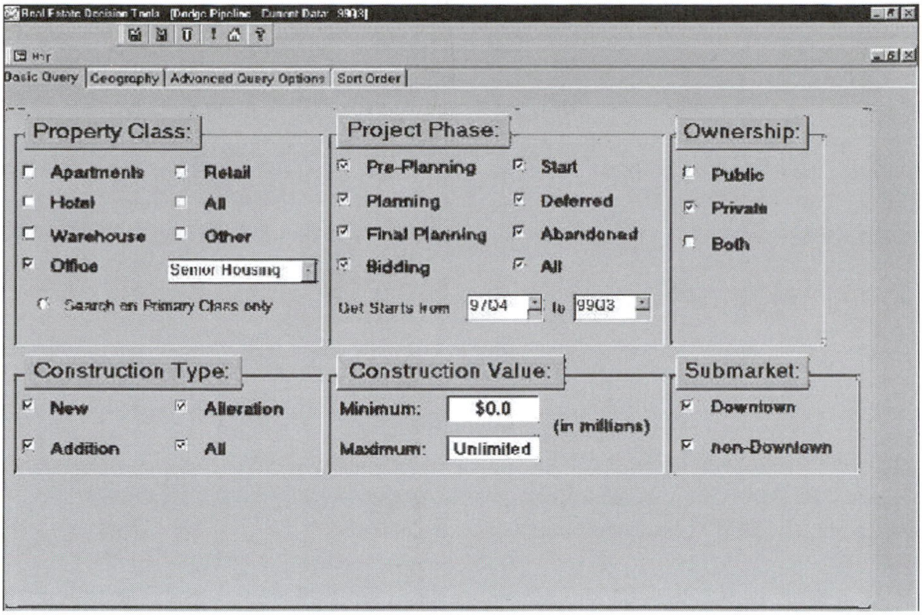

Figure 6.2 Choose product type, F. W. Dodge pipeline.

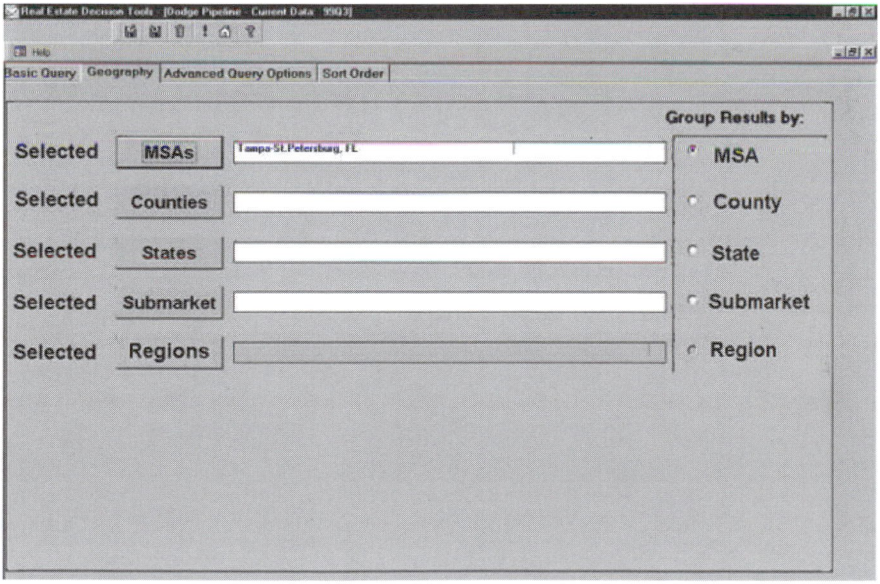

Figure 6.3 Select trade area, F. W. Dodge pipeline.

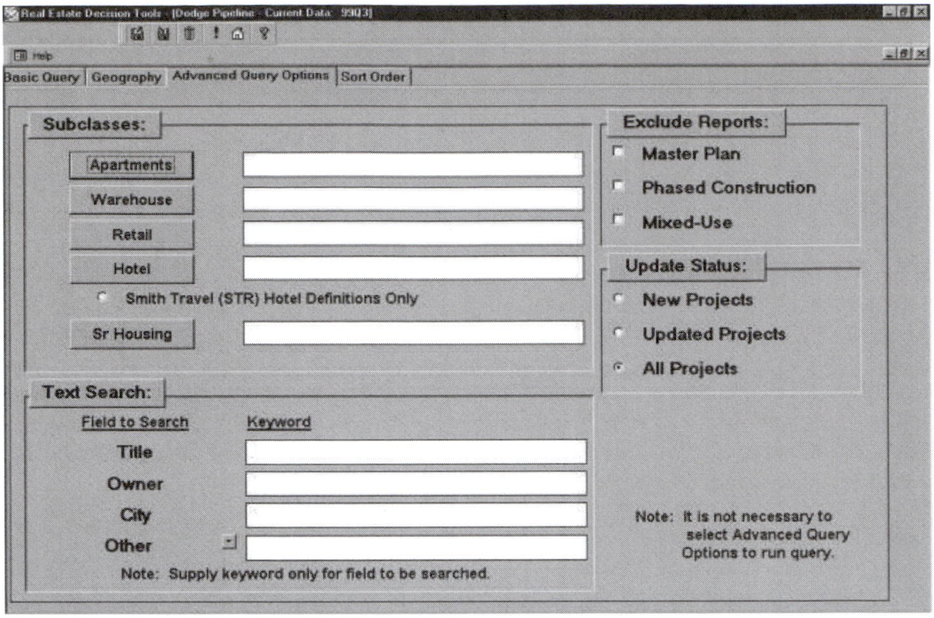

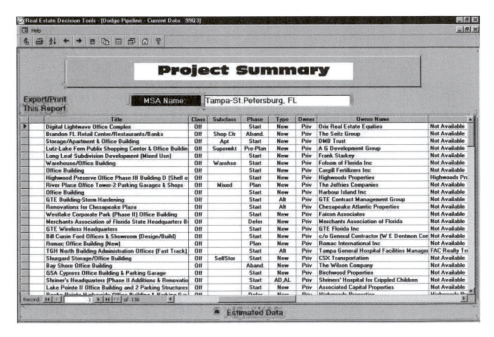

Figure 6.4 Refine search for competitive product, F. W. Dodge pipeline.

Title	Class	Subclass	Phase	Type	Owner	Owner Name	
Digital Lightwave Office Complex	Off		Start	New	Priv	Orix Real Estate Equities	Not Available
Brandon FL Retail Center/Restaurants/Banks	Off	Shop Ctr	Aband.	New	Priv	The Seitz Group	Not Available
Storage/Apartment & Office Building	Off	Apt	Start	New	Priv	DMB Trust	Not Available
Lutz-Lake Fern Publix Shopping Center & Office Buildin	Off	Supermkt	Pre-Plan	New	Priv	A G Development Group	Not Available
Long Leaf Subdivision Development (Mixed Use)	Off		Start	New	Priv	Frank Starkey	Not Available
Warehouse/Office Building	Off	Warehse	Start	New	Priv	Folsom of Florida Inc	Not Available
Office Building	Off		Start	New	Priv	Cargill Fertilizers Inc	Not Available
Highwood Preserve Office Phase III Building D (Shell o	Off		Start	New	Priv	Highwoods Properties	Highwoods Pr
River Place Office Tower-2 Parking Garages & Shops	Off	Mixed	Plan	New	Priv	The Jelfries Companies	Not Available
Office Building	Off		Start	New	Priv	Harbour Island Inc	Not Available
GTE Building-Storm Hardening	Off		Start	Alt	Priv	GTE Contract Management Group	Not Available
Renovations for Chesapeake Plaza	Off		Start	Alt	Priv	Chesapeake Atlantic Properties	Not Available
Westlake Corporate Park (Phase II) Office Building	Off		Start	New	Priv	Faison Associates	Not Available
Merchants Association of Florida State Headquarters B	Off		Defer	New	Priv	Merchants Association of Florida	Not Available
GTE Wireless Headquarters	Off		Start	New	Priv	GTE Florida Inc	Not Available
Bill Currie Ford Offices & Showroom (Design/Build)	Off		Start	New	Priv	c/o General Contractor (W E Dentmon Con	Not Available
Romac Office Building (New)	Off		Plan	New	Priv	Romac International Inc	Not Available
TGH North Building Administration Offices (Fast Track)	Off		Start	Alt	Priv	Tampa General Hospital Facilities Manager	FAC Realty Tr
Shurgard Storage/Office Building	Off	SelfStor	Start	New	Priv	CSX Transportation	Not Available
Bay Shore Office Building	Off		Aband.	New	Priv	The Wilson Company	Not Available
GSA Cypress Office Building & Parking Garage	Off		Start	New	Priv	Birchwood Properties	Not Available
Shriner's Headquarters (Phase II Additions & Renovatio	Off		Start	AD AL	Priv	Shriners' Hospital for Crippled Children	Not Available
Lake Pointe II Office Building and 2 Parking Structures	Off		Start	New	Priv	Associated Capital Properties	Not Available

Figure 6.5 Sample display of F. W. Dodge selected properties in the pipleline. All data fields can be exported to spreadsheet or GIS software.

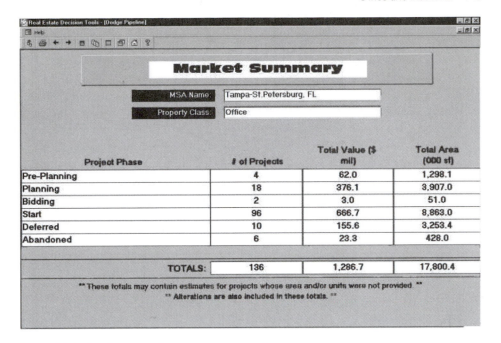

Figure 6.6 F. W. Dodge pipeline summary for trade area.

that of the actual trade area, then the automatically generated report of figure 6.6 will be a good competitive summary in the primary trade area for the proposed project. Otherwise, the same categories of summary measurements can be derived in the GIS software following appropriate spatial selection and GIS software specific procedures.

Finally, import the .xls or .dbf file into the GIS software (figure 6.7). All the data fields available in the F. W. Dodge database remain available in the GIS software. The GIS software has capabilities for structured query language-like selection as well as spatial selection. Spatial selection may include a refined trade area boundary. The GIS software allows the analyst to visualize the location of the competitive projects and to derive geographic summary measurements such as quantity of competitive projects in the pipeline by square footage by census tracts.

Deriving Inventory with Comps.com

As with pipeline data, deriving the current inventory of commercial property by product category, for the larger U.S. markets, is a point-and-click operation on a desktop computer. The business geographer then can place the data into GIS software for subsequent submarket analysis. The following is a demonstration of deriving inventory data with the Comps.com product.

First select the product type that would be considered a competitive product for the proposed project. The Comps.com options are shown in figure 6.8. The user may choose between major categories of apartment, commercial land, hotel/motel, industrial land, industrial, mobile home park, office, residential land, retail, and other com-

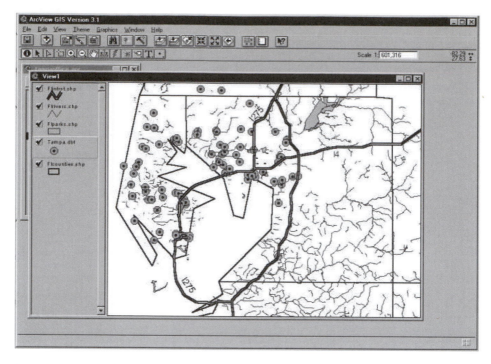

Figure 6.7 F. W. Dodge pipeline data visually displayed in a GIS software program and available for further geographic analysis.

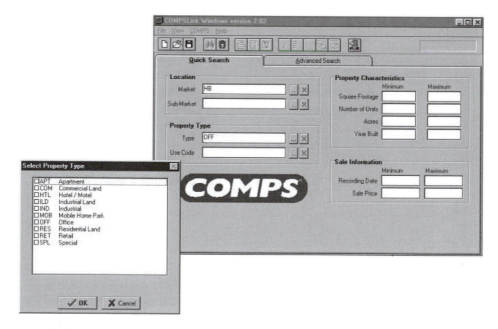

Figure 6.8 Comps.com inventory supply data for commercial real estate. HB denotes Hillsborough County, Florida.

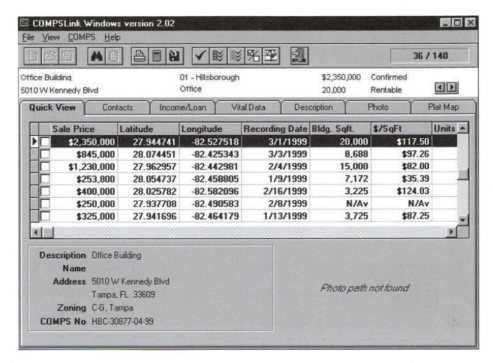

Figure 6.9 Sample results from Hillsborough County office query in Comps.com.

mercial properties. Within each large product category, Comps.com has additional categories that the analyst may choose from to refine a real estate product category considered competitive with the proposed project.

Second, the general trade area is selected. Hillsborough County, Florida, is used for the illustration. The software form in figure 6.8 allows the analyst to identify the trade area from which data will be selected; the user may choose from categories of MSAs or counties. Third, the query is run on the Comps.com data, and 168 office properties in Hillsborough County are selected. Figure 6.9 shows a sample of the individual records. Figures 6.10 and 6.11 show detail on the property highlighted in figure 6.9, including property tax, expenses, vacancy rate, net income, cap rate, and developed square footage. Figure 6.12 shows a photo of the property. Figure 6.13 shows a plat map of the property. Each record may be viewed within the Comps.com software as shown in figures 6.10–6.13. All records may be exported for later analysis in spreadsheet or GIS software in the same manner as described above for the F. W. Dodge pipeline data.

Example GIS and *Black's Guide to Office Space* Inventory

Black's Guide to Office Space (1997) lists 311 office buildings within Hillsborough County. *Black's Guide* reported 23,375,886 total square feet of office space, of which 3,279,452 square feet were vacant, yielding a 14.03 percent vacancy rate. When this

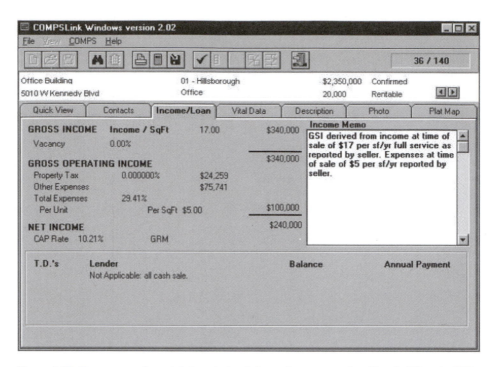

Figure 6.10 Comps.com financial descriptive information on sample office building in Hillsborough County, Florida.

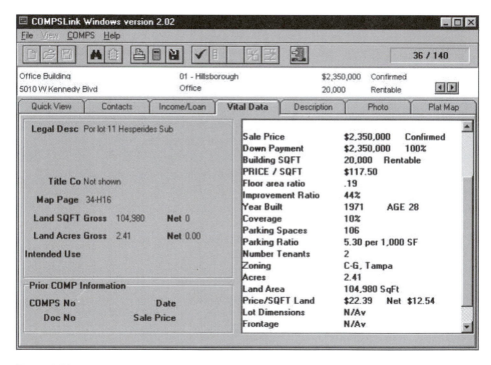

Figure 6.11 Comps.com descriptive facility detail of sample office building in Hillsborough County, Florida.

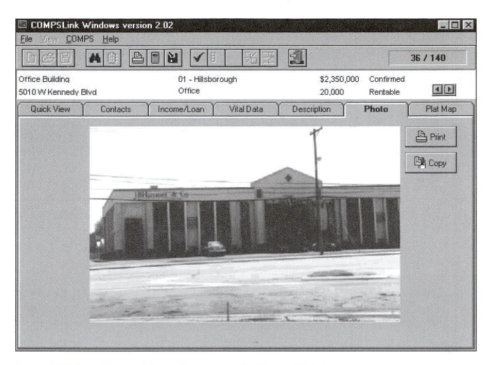

Figure 6.12 Sample image from Comps.com database of Hillsborough County, Florida, office buildings.

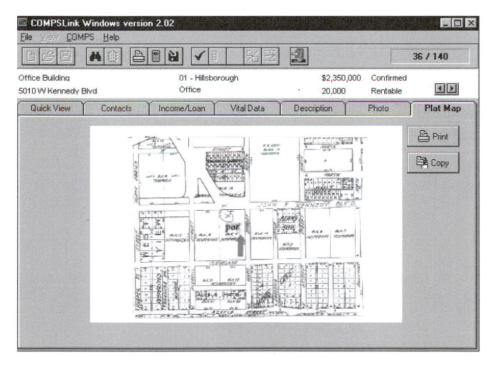

Figure 6.13 Plat map of sample property from Comps.com data base of office buildings in Hillsborough County, Florida.

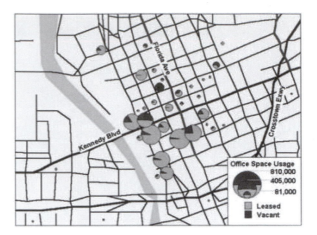

Figure 6.14 Map of supply of office space created with GIS (from Thrall and Amos 1999b).

example was created, one building was under construction, which would add 90,000 square feet to the total office space in the county. *Black's Guide* does not include latitude and longitude coordinates, so the business geographer needs to assign coordinates to the office buildings. *Black's Guide* does include addresses for the office buildings. Desktop GIS software can assign a latitude and longitude coordinate based on an address.

In addition to displaying the location and information of the properties within the database, GIS can display and analyze relationships of the properties based on the attribute data within the property database. GIS allows the market analyst to create thematic maps of the office buildings within submarkets that display the total amount of space in each building, the space leased, and the space available. Full-featured desktop GIS software can display pie charts at the location of observation, where the divisions within the pie chart can be directed to display proportional segments of a data attribute. Geopositioned pie charts can be used to simultaneously display the amount of space rented in a building or submarket and the space vacant in the building or submarket, while the radius of the pie chart can be set proportional to the total square footage of the building or submarket. GIS brings the ability to simultaneously visualize multiple fields of information within a spatial context.

The location of the pie charts in figure 6.14 shows the geographic position of the office buildings within the Tampa CBD submarket. The wedges of the pie charts show the quantity of space rented within the office building and the quantity of space vacant within each office building.

The preceding GIS-based procedure for visualization of office submarkets was used to define the Hillsborough County office submarkets. *Black's Guide*, which also provides geographic boundaries for office submarkets, was referenced before the boundaries were drawn. Six separate submarkets were specified: Westshore, downtown, northeast, northwest, southeast, and southwest, as shown in figure 6.1.

A GIS point-in-polygon procedure can aggregate selected attribute values within each of the office submarkets. Figure 6.15 displays a graph of the total office space, leased office space, and the vacant space by submarket for Hillsborough County. The

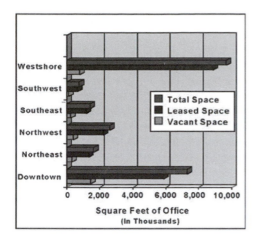

Figure 6.15 GIS-derived supply of office space (from Thrall and Amos 1999b).

GIS calculations revealed that the Westshore submarket, with 9,650,203 square feet of office space, was the largest submarket within Hillsborough County. The Westshore submarket was calculated to have 779,564 vacant square feet. The 8.08 percent vacancy rate was the lowest of any of the Hillsborough County submarkets.

The downtown submarket was calculated as being the second largest submarket in total office space, with 7,316,426 square feet. The vacancy rate for the downtown submarket was 19.86 percent. Further disaggregation of offices into classifications that are standard for the industry revealed that 75 percent of the downtown submarket is considered class A, and class A office property had a 9 percent vacancy rate.

Lower grade class B space may at one time have been considered class A, but might be older construction that has not been upgraded to standards expected of newly built office space. In downtown Tampa, as is typical in most markets, the lower the grade of office space, the higher the rate of vacancy.

In Hillsborough County, the northwest submarket was calculated as the next largest submarket in terms of square footage, followed by the northeast, southeast, and southwest submarkets. The aggregate vacancy rates across all classes of office property for the four suburban submarkets was calculated as: northeast, 18.83 percent; northwest, 14.99 percent; southeast, 16.48 percent; and southwest, 15.40 percent. So the vacancy rate in the Tampa CBD was calculated as being the highest in Hillsborough County. Office space users within Hillsborough County therefore have revealed a preference to rent office space outside of the traditional boundaries of the Tampa CBD within the period of this example.

Step 3: Demand Analysis

The demand for office space comes from firms seeking quarters for their clerical, supervisory, and management personnel. The driving force behind the demand for new office space is largely an increase in white-collar employment (see Shilton and Webb 1991). Positive influences on local job growth arise from local firms upgrading and expanding existing operations, the creation of new local firms, and outside firms relocating to the local market. Negative influences resulting in a reduction in the demand for office space within a submarket arise from local firms relocating out of the submarket, firms closing their operations, reducing the scale of their activities, or reducing their need for space because of alternative office space strategies such as telecommuting.

There are four steps to demand analysis:

- Estimate white-collar employment for a chosen base year, and the projected number of white-collar employees in a future or target year
- Compute the square footage of office space per white-collar employee
- Compute the growth of white-collar employees by subtracting the number of white-collar employees for the base year from the projected
- Forecast the additional quantity of office space required in the target year using calculations from the preceding three items.

There are a variety of issues associated with the correct calculations of each of the above items. For example, white-collar employment must be forecast from the base year to the target year. Greenberg (1978) demonstrated techniques to forecast local population and employment projections. Greenberg's proportionate share analysis assumes that past trends in local employment relative to employment at the state level will persist into the near future. In proportionate share analysis, the ratio of local employment to state employment is calculated for a base year. Proportionate share analysis assumes that the locality will continue to grow in proportion to the state. County and/or metropolitan figures are available from a variety of governmental offices, including the local government and various federal agencies. However, the analyst should keep in mind that the data projections are likely to have been made using the same proportionate share analysis. Many data value-added resellers sell forecasts at the state, county, and metropolitan-area levels, and even for smaller geographic units. The analyst should keep in mind that the smaller the geographic unit and the farther ahead the time horizon, the more likely the forecast will deviate from the true value.

The ratio of local employment to state employment is multiplied by an estimate of state employment in the target future year. An estimate of future county employment can be used if a reliable measure is available. The map in figure 6.16 shows the white-collar employment by census tract from the 1990 U.S. Census of Population for Hillsborough County.

Table 6.2 shows the employment of Hillsborough County. The U.S. Census of Population classifies the population by type of occupation. The white-collar employment rate was created using the following fields from the 1990 U.S. Census of Population: executive, administrative, and managerial; professional specialty occupations; technicians and related support occupations; sales occupations; and administrative support occupations. The analyst may use his or her judgment to decide which occupational category to include. For example, in the product category being analyzed, technicians may not be relevant for some office space; they instead may be situated in industrial space. Likewise, sales occupations may be situated in retail space.

For the demonstration calculations, 1996 was chosen as the base year and 2005 as the target year. The percentage of each type of occupation is derived by dividing each

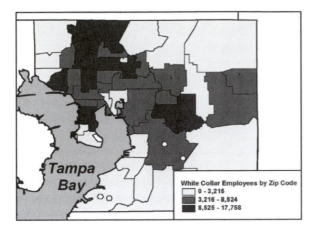

Figure 6.16 White-collar employment, 1990, by census block group in Hillsborough County, Florida (from Thrall and Amos 1999b).

classification of white-collar workers by the total number of persons employed in 1990. The 1996 estimate of total persons and employed persons was obtained from *Regional Economic Growth in the United States: Projections for 1997–2025* (Terleckyj and Coleman 1997). The 1996 estimate of employment by white-collar occupation was then calculated by multiplying the 1990 percentage value by the 1996 estimated employed persons for Hillsborough County according to Terleckyj and Coleman (1997).

The geographic technology industry has responded to the need for this type of data projection. Data from this segment of the geographic technology industry can save the analyst the time required to perform calculations like those shown in table 6.2. The analyst should keep in mind that the data vendor may have calculated the projections in a manner similar to that outlined above. In other words, a variety of judgmental decisions and assumptions by the data creator are embedded in the projection and may or may not correspond to the same decisions that the analyst would make if the data were self-generated.

The next step in the demand analysis is to compute the number of square feet per employee. Estimating the amount of office space used per employee by SIC code can refine the square footage per worker measurement. The example calculations were

Table 6.2. Estimation of 1996 and 2005 employment by white-collar category, Hillsborough County, Florida

	Employed persons	White-collar employed persons	Executive administrative and managerial	Professional specialty occupations	Technicians and related	Sales occupations	Administrative support occupations
1990 Census	412,188	256,678	54,283	53,243	16,003	58,659	74,490
Percentage			13.17	12.92	3.88	14.23	18.07
1996 Estimate	610,900	380,420	80,452	78,910	23,717	86,937	110,400
2005 Estimate	708,400	441,135	93,293	91,505	27,503	100,813	128,021

From Thrall and Amos 1999b.

performed without consideration of the variation in office space by occupation; instead, they calculated the space per employee as the total occupied square footage of office space divided by total white-collar employment:

$$\frac{20{,}096{,}434 \text{ ft}^2 \text{ occupied office space}}{380{,}420 \text{ white-collar employees}} = 53 \text{ ft}^2 \text{ per employee}$$

One of the continuing debates among analysts is how much square footage is required for each employee. The above calculation of 53 square feet per employee corresponds to the square footage absorbed in Chicago in 1995, as reported by the Urban Land Institute (Gause et al. 1998, p. 39). The ULI also reports that the office space allocation per employee for the nation during the 1990s was in the range of 200–250 square feet. The evidence is, as discussed in chapter 2, that the practice of hoteling office space is reducing the office space needed per employee. Also, more space-efficient architectural designs have reduced the space per employee. Therefore, 53 square foot per employee is used for this demonstration, as it arises out of the above projections, but to some analysts this figure might be considered to be too little or too much office space for the twenty-first-century employee.

The growth of white-collar employees within an office market is computed by subtracting the estimated 1996 white-collar employment in the base year from the projected white-collar employment for the target year of 2005:

$$441{,}135 \text{ (projected)} -380{,}420 \text{ (estimated)} = 60{,}715 \text{ (projected increase)}$$

The last step of the demand analysis is to forecast the quantity of office space required in the future target year. The projection is calculated as the estimated increase in number of white-collar employees times the estimated square footage of office space per employee:

$$60{,}715 \text{ (projected increase)} \times 53 \text{ ft}^2 \text{ employee} = $$
$$3{,}217{,}895 \text{ (increase in demand)}$$

The increase of 60,715 white-collar employees in the county provides an estimated total need for 3,217,895 gross square feet of office space over the 10-year projection period. Prorated by year, there is expected to be an average yearly increase in market demand at the countywide level of 321,790 square feet between 1996 and 2005.

The proportionate share analysis can be extended using GIS to spatially disaggregate the countywide measurements by office submarket. This approach naturally assumes that the submarkets will grow at historical proportions to the rest of the county. The Westshore submarket contains 44.14 percent of all leased office space in Hillsborough County, calculated as the ratio of Westshore leased office space (9,650,203 − 779,564) to Hillsborough leased office space (23,375,886 − 3,279,452). If the Westshore submarket can continue to claim a 44.14 percent market share of leased office space, then Westshore will have an increase in office space demand of 1,420,390 (44.14% × 3,217,895) square feet. The analysis should include best case and worst case scenarios allowing a range for future market share captured by the submarket. Part of that increase in demand can be met by the present 779,564 vacant square feet, meaning that in Westshore, 654,739 square feet of office space can be built and rented (absorbed) over the next 10 years, averaging out to about 65,474 of office square footage per year.

Office Space User Chooses a Location

Early in this chapter, an analogy was made between office employees and retail customers. That analogy can be put to valuable use when an office space user considers between relocating to another office submarket and remaining in their current office submarket. A decision to evaluate a relocation of office space may arise because of significant expansion or contraction of operations or because of some dissatisfaction with the current office submarket or office building. Text box 6.3. lists a five-step procedure that an office user would go through to evaluate possible sites or submarkets in relocating existing office operations. Among the significant costs confronting the office space user is employee dissatisfaction with the move because of increased commute time, and consequently problems of employee retention.

Step 4: Final Report Including Absorption Analysis

Absorption analysis integrates the work on trade area, demand, and supply. Absorption analysis is generally integrated into the final report. The final report to the client presents information in such a manner as to support the recommendation for a particular decision.

Office market absorption is the quantity of office space that an office market can be expected to rent when new supply is introduced. The ability of an office market to capture a portion of the aggregate local demand will depend on the marketability of the region. The absorption rate is often evaluated by tracking recent absorption trends for the region.

Hillsborough County has a supply of 23,375,886 square feet of office space, of which 20,096,434 is currently leased. An additional 90,000 square feet in one building is under construction. The current available supply is 3,279,452 square feet. The growth in white-collar employment by 2005 will require 3,217,895 of office space. The total existing supply of 3,279,452 square feet of office space constitutes more than a 10-year supply of office space if it is absorbed at a rate of 321,790 per year. The absorption rate among the submarkets within Hillsborough County presents a different story. The Westshore submarket has the largest quantity of office space and the lowest vacancy rate among all the submarkets. In Thrall and Amos's (1999b) GIS evaluation of the Hillsborough County office market, the Westshore submarket represents the best opportunity for office market development in that county.

Based on the above analysis, the report would indicate that, at the countywide level (Thrall and Amos, 1999b), there will not be a need for additional space over the next ten years. However, the analysis at the submarket level indicates that development opportunities exist in at least one or more of the submarkets. Specifically, for the Westshore submarket in Tampa, the market could absorb new office space.

Box 6.3. Five steps to identify prospective new office sites and sub-markets*

1. Perform a needs assessment. A needs assessment requires personal interviews. The analyst should perform the interviews, as he or she does not have a stake in a particular locational outcome. The interviewer hears from a stratified random sample of employees at each rank and level within the organization using the office space what they consider to be important in terms of the need for new office space and what criteria should be used to choose the location of a new facility. *Modified Delphi* is another term that has at times been used for the same step. For further discussion, see Thrall and McCartney (1991).

2. Perform appropriate analysis based on criteria identified in step 1. GIS software in combination with appropriate data are normally used to reveal prospective locations. Some of the criteria might be straightforward, thematic-like mapping, such as crime rate not exceeding a certain level or availability of housing in a specified price range within a given distance from the new facility. Other criteria might require more complex geographic analysis that assigns geographic coordinates to all the residences of the affected employees and calculates their change in commute time or commute distance between the old facility and a proposed new office facility. The corporation's division of human resources might propose criteria for choosing a location where the geodemographic information suggests that there is a supply of particular kinds of labor nearby. Other possible criteria might include the calculation of the impact on dual income households and ride sharing. See main text for further discussion.

3. Identify prospective locations based on the first and second steps. GIS software in combination with appropriate data and analysis serves to narrow the list of prospective suitable locations.

4. Visit the locations shown to be acceptable in step 3. The analyst must combine the above analysis with his or her own qualitative evaluation of the locations. Site visits are important because not all relevant criteria and phenomena can be quantified. The analyst's qualitative evaluation of the sites is based on his or her knowledge and experience with business geography, urban geography, real estate, market analysis, and general site selection. The analyst then integrates this qualitative assessment with local expert knowledge.

5. Write a final report of the findings of the above steps. In addition to input from the employees as referenced in the needs assessment of step 1, throughout the work, the analyst should integrate the concerns and desires of the corporation's division of human resources and division of real estate. Today, there is a separation between the roles of the business geographer site analyst and the expert in real estate. Just as a business geographer/real estate market analyst would not be involved in hiring or firing decisions of human resources, the analyst should also not attempt to take over the existing role of the division of real estate in their contract negotiation or legal issues dealing with purchase or lease. However, as those divisions request input from the analyst, the analyst should be willing to provide whatever guidance might be relevant and related to his or her expertise.

*Cynics may comment that the market analyst should begin by determining where the CEO would like to live and then choose a location that is best suited for the CEO's commute to work.

Industrial, R&D, and Flex Space: A Variation on Office Space Analysis

The methods introduced in this chapter for market analysis of office space apply, with some modification, to industrial space research and development (R&D) and flex (flexible use) space as well. Demand for office space can be the outgrowth of business activities in the local, regional, national, or global economy. However, unlike office space, which can have a significant component that arises from the local economy, demand for industrial space is instead the result of larger national and global economic forces (see chapter 2 on economic base.) A large literature has been developed for 150 years in the field of economic geography on why industry locates where it does. The basic principles in that literature allow the analyst to evaluate the local prospects for an increase or decrease in local industrial space demand. The basic principles are necessary background to modify the foregoing office market analysis for industrial space, R&D and flex space.

Why Industry Locates Where It Does

The reasons for industrial location can be grouped into seven basic categories. An analyst advising a manufacturing firm on where to locate new facilities or relocate existing facilities will evaluate how each of the seven categories relate to the particular firm and will evaluate how the local market corresponds to the needs of the industrial firm. An analyst evaluating the prospects for future industrial development at a site or market area situation will also consider the same seven categories in the assessment.

Following is a brief overview of each of the seven categories. The seven categories are presented in rank order of importance for a typical manufacturing firm, with the most important presented first. However, conditions might lead a particular firm to place a different emphasis on each of the seven categories.

Reason 1: Agglomeration Economies

The most important reason for most manufacturing firms to locate where they do is *agglomeration economies*—the benefits associated from being geographically situated near other manufacturing, even competitive manufacturing, that require similar factors of production, including labor, public infrastructure, private support services, and so on. Agglomeration comes in two forms: economies of scale and information.

Typically, the greater the output, the lower the average cost of production per unit of output. However, as congestion sets in, the average cost may increase. This describes the U-shaped average cost curve faced by a firm. Similar to the analogy of a firm, a region might have similar cost characteristics. As production begins in a region, the average cost of producing in the region may initially be high. As production increases, the average cost might decline. Ultimately, congestion sets in, so the regional average cost of production rises.

To the extent that regional average costs are internalized to the individual manufacturing firm, manufacturing in a region where other firms have not already established a foothold might be more costly to the firm than locating where other manu-

Table 6.3. Illustrative check list for conformity of manufacturing firms' needs for agglomeration economies to local situation

Importance to the firm	Description of the local market area (situation)
Manufacturing process requires inputs available locally, and without the demand by other firms the providers of the inputs will not have the threshold demand to remain in business.	Does sufficient local demand exist for required providers of inputs and services to thrive?
Manufacturing process requires public facilities to be available locally, such as sewer and electrical capacity.	Does necessary capacity exist?
Is it necessary to be part of a frequent face-to-face dialog or chance encounter with other firms, institutions, or organizations, to remain competitive?	Which firms are already located nearby, and what type of synergy arises because of the geographic density of those firms?

facturing firms are already operating. In other words, a firm might minimize its costs by locating where other firms, even competitive firms, already exist. This is one argument for manufacturing firms to agglomerate. The result is that the more circumstances have been a particular way in the past, the more likely they will remain that way in the future.

Agglomeration also can be viewed from an informational standpoint. Goods that are part of a fast-paced and changing industry, where styles of the finished good are seasonal tend to locate together. The decision makers need to be part of the information loop so that they can anticipate the quickly changing economy. Designer clothing and the high-tech software industry fall into this category. Table 6.3 is a checklist for identifying what is important to an industrial firm and how the proposed site and situation correspond to the requirements of the firm.

Reason 2: Variation in Manufacturing Costs

Industrial location decisions are made with the objective of minimizing costs of manufacturing (see table 6.4). The larger the fixed cost of the manufacturing facility, the greater will be the expected length of time required for the firm to amortize its investment in the location. Hence, large costs associated with manufacturing at the location need to be projected over a long time horizon. The costs in rank order that a typical manufacturing firm must consider are discussed in this section.

Labor   The second most important reason an industrial firm locates where it does is variation in manufacturing costs. And the most important variation in manufacturing cost facing most firms is the cost of labor.

In the post–World War II years, labor costs for the typical U.S. firm ranged between 40 percent and 60 percent of the total cost of manufacturing. Today that number remains about the same. However, the structural makeup of that cost has changed. In the 1950s, about 75 percent of the labor cost was in production labor and 25 percent in administration, the white-and pink-collar jobs of the era. Those percentages have now switched for many North American firms.

Many firms heavily dependent on low-skilled production labor have left the United States for other nations, including countries in Latin America and Asia. Since the 1950s, there has been a steady substitution of capital for labor, thereby automating many manufacturing processes. There has also been a structural shift in the U.S. economy to more skilled, high-tech enterprises.

The change to highly skilled technical workers has made the twenty-first-century industrial location decision different from the twentieth- and nineteenth-century industrial location decision. But in terms of labor, the general objective remains the same: locate where the supply of labor with the necessary skills will not result in labor shortages and where the cost of employing labor with the required skills is projected to be least over the relevant time horizon.

The availability of highly skilled labor is positively correlated with public expenditures in education, from childhood through advanced graduate levels. The greater the expenditure at the per-capita and per-pupil levels, the greater the quality and the supply of skilled labor. States and localities that have a history of sustained high levels of expenditures in education are in a more competitive position to attract manufacturing dependent on skilled, knowledgeable labor.

Labor supply is also affected by in and out migration, as discussed in chapter 2. As a household's income rises above the median level, the household is willing to trade greater percentages of additional income earned for acquisition of urban-built amenities and natural environmental amenities. In other words, the greater the amenities, the greater the supply and the lower the price of highly educated and highly skilled labor. People are willing to sacrifice to live in a high-amenity city, including receiving lower wages and paying higher housing costs. A manufacturing firm might need to pay a premium for highly educated and highly skilled labor if the facility is located in a decaying, nineteenth-century, polluted industrial city.

What are amenity cities? High-amenity cities are surrounded by a scenic natural environment, devoid of urban sprawl and devoid of intensive human development. The high-amenity city is one that is well planned, often with a thriving historic CBD, as well as attractively designed regional retail centers and housing subdivisions. The regional shopping centers might have been designed as attractive "lifestyle centers." The housing subdivisions might have been designed to conform to notions of sustainable urban development. Examples of high-amenity cities in the United States include Portland, Oregon; Boulder, Colorado; and Santa Fe, New Mexico. In Canada, Toronto would be considered a high-amenity city, indicating that it is not only small and intermediate-sized cities in areas bestowed with natural amenities and favorable climate that can lay claim to this category. A commonality between high-amenity cities is their urban planning.

The changing structure of the household of the twenty-first century also affects the supply of labor. Some estimates are that more than 60 percent of working-age women under 40 years of age are actively engaged in occupations outside the home. The result is that relocation decision by an employee is no longer made merely to maximize the career opportunities of one individual, but instead to maximize the career opportunities of both partners in the household. The household will make a location decision that maximizes their joint welfare. Some communities do not have sufficient scope or breadth of economic base to allow for two diversified and highly-specialized occupations. The household may base its location decision on the question, which city can we both reside in and both be able to work in our chosen occupations? A career move to a small town might appeal to one member of the household, and the small town might offer great amenities; however, the small town might not have the scope of local economy to allow the other significant member of the household to engage in the profession they choose. Thus, the analyst must also consider the scope and breadth of the local economy.

Capital   Generally, the larger the city, the greater the availability of venture capital. Venture capital is a necessary requirement for most new firms. The price of venture capital can vary from region to region. Venture capitalists often require that those firms to which they lend be located in their city. The venture capitalist wants to look out the office window and see "smoke coming out of the smokestacks" of the firm they have lent to. New York City and San Francisco are examples of cities particularly noted for availability of venture capital.

Land   The twenty-first-century industrial firm generally consumes more land than the nineteenth-century firm, as explained in chapter 2. Within a city, because land values decline toward the urban periphery, and because of the availability of large parcels of land, firms choosing new locations tend to be biased toward the urban periphery.

Land prices and availability are also important between cities. As expected from the general theory (see chapter 3), the price of acreage is much greater in large Los Angeles than in small Gainesville, by a multiple of about ten. Even within a region, land prices can be significantly different. The price of acreage is greater in larger Orlando than in smaller Gainesville. The two cities are 100 miles apart, and land prices for parcels in comparable situations in Orlando are greater by a multiple of about three. A land-intensive warehousing operation might find it attractive to locate at a somewhat distant center and perhaps have higher transportation expenditures than pay significantly higher land costs.

Land costs also affect the supply of labor. The greater the land costs, the greater the cost of housing. If compelling amenities do not offset housing costs, then a labor shortage might ensue.

Heating and Air-Conditioning   The U.S. South has an advantage over northern Midwest and northeastern locations in that the naturally occurring temperature range for some industrial operations can be had without expensive heating. The U.S. South is at a disadvantage in that during the summer months the humidity and temperatures combine to make for what to many is an intolerable climate without air conditioning. It has frequently been said that without air conditioning, the South would not have economically developed. The analyst must consider the importance of the expense of heating and air conditioning to the manufacturing process, its cost locally, and how that cost compares to other competitive locations.

Energy   Energy is placed in a separate category from heating and air conditioning to distinguish the sources of energy for manufacturing from that of only climate control. Some manufacturing facilities are great consumers of energy, representing an expense

Table 6.4. Illustrative checklist for importance of variation in manufacturing costs and conformity to the local market situation

Element	Importance to the firm	Description of the local market area (situation)
Labor	What is the amount of (a) Unskilled production labor required now and in the future? (b) Skilled technical labor required now and in the future? (c) Highly educated managerial labor required now and in the future?	(a) Unskilled production labor available now and in the future? (b) Skilled technical labor available now and in the future? (c) Highly educated managerial labor available now and in the future?
Capital	Does the firm require (a) High-risk venture capital? (b) Frequent bridge loans?	(a) Availability of venture capital in market? (b) Cost of venture capital in market as compared to other markets? (c) Availability and cost of bridge loans in local market of the amount that firm requires?
Land	What is the (a) Amount of land that the firm requires for its operations? (b) Budget allocation firm is willing to allocate to land acquisition?	(a) Availability of land for present development and future expansion? (b) Cost of land for present development and future expansion?
Heating and cooling	(a) Is the manufacturing process limited to a specific temperature range? (b) What is the budget allocation the firm is willing to allocate to heating and cooling?	(a) Can the required temperature range be met by natural environmental conditions or must the temperature range be met by construction, and if so at what cost? (b) What is the local price for heating and cooling?
Energy	(a) What are the energy needs of the firm? (b) What is the budget allocation the firm is willing to allocate to energy?	(a) Can the required energy needs be met by local infrastructure already in place? (b) Price for energy consumed at that site?
Waste disposal	(a) Does the manufacturing process result in nontoxic or toxic waste? (b) What is the budget allocation the firm is willing to allocate to toxic or toxic waste disposal?	(a) Does the locality offer industrial nontoxic or toxic waste disposal? (b) What is the cost of industrial nontoxic or toxic waste disposal?
Taxes	(a) Does the manufacturing firm maintain an operation that is particularly susceptible to local taxation? (b) Are taxes a high percentage of the total operations of the business?	(a) Does the locality have in place taxes that target a particular type of operation? (b) Are the taxes of the locality particularly high as compared to other places that may be competitive for the same type of manufacturing facilities?

even greater than labor. Aluminum smelting is one such industry that locates in regions served by inexpensive hydroelectric power. The analyst needs to consider the supply of available energy for manufacturing use and the cost in comparison to competitive locations.

Waste Disposal   In the nineteenth century, little attention was paid to disposing of industrial waste. In the twenty-first century, the hazard of improper waste disposal to human health is more understood and more regulated. Both toxic and nontoxic waste can be a byproduct of manufacturing processes. Waste disposal can be expensive, but more so in some localities than in others. For a waste disposal facility to receive federal certification to receive toxic waste, a variety of physical geographic conditions must be met.

The physical geography of some regions simply does not match those conditions that would allow for federal certification. Florida can be described as one of the largest rivers in the world, moving from the Florida–Georgia border to the Everglades in southern Florida. Unlike other rivers, only a small part of the waterways of Florida are visible at the surface; most are underground. Whatever is buried moves down river to some other community's aquifer for drinking. The result is that toxic waste generated in Florida must be shipped great distances, at significant expense. The analyst must evaluate the quantity of waste generated by the firm or industry, the availability of toxic waste facilities, and the cost of disposal.

Taxes   It may come as a surprise to many to have taxes listed as the last and most insignificant item in this list. It is listed last because within a nation, the variation in taxes from one location to another is seldom significant. The smaller the variation from one place to another, the less important that element is in the location decision. Geographically, taxes tend to be a fixed cost, thereby not affecting the location decision. Any variation from place to place that does exist is normally overwhelmed by the locational variation in other costs, such as labor. Does a firm make a location decision based on that large labor cost percentage, or the comparatively small differential in tax percentage? Why then do we hear more about taxes than any of the other items on this list?

Taxes are subject to legislative change. A vote can change taxes, nearly immediately. Instead, a vote to build up a supply of highly educated labor requires a generation or more to witness the benefits. Politicians want to appear in control during their term of office. Taxes are something that can be controlled during that term. Because most politicians have the same objective, then a spatial equilibrium of taxation arises, and seldom is one community out of balance with another community. Geographic differences in taxation are rarely important in a location decision. Nevertheless, the business geographer must perform due diligence and document local taxes.

Reason 3: Resources

Generally, if by way of the manufacturing process there is a significant decrease in weight, bulk, fragility, perishability, or hazard, then the manufacturing firm will locate where the resources are. If there is a significant increase in any of those items, the firm will be biased to locate near the market where the final product is consumed.

Table 6.5. Illustrative checklist for importance of resources and conformity to the local market situation

Importance to the firm	Description of the local market area (situation)
Does the manufacturing process result in an increase in weight, bulk, fragility, perishability, or hazard?	The manufacturing firm will be biased to locate where access to its market is best, and where transportation cost of the finished product is least
Does the manufacturing process result in an decrease in weight, bulk, fragility, perishability, or hazard?	The manufacturing firm will be biased to locate where access to its raw material inputs is best, and where transportation cost of the raw material to the manufacturing facility is least

Copper smelters are located at the mine head because of the loss of weight achieved in the manufacturing process. Newspapers have been produced in the market where they are consumed because news is highly perishable. Fish canneries are located near the port of the fishing fleet for the same reason. Japanese and German automobile assembly plants are located in the United States because once a car is assembled it is bulky, fragile, and expensive to transport. Because two-thirds of the U.S. population live within a day's drive of the Appalachian Mountains, then so too are the Japanese and German assembly plants located within a day's drive of Appalachia. Cotton was once king in the southern U.S. states. Cotton before it is processed is very bulky. Therefore, to minimize transportation costs, textile mills dotted the South to be near the source of inputs. The boll weevil killed king cotton. Because the input vanished, the textile mill industry also declined.

Table 6.5 is an analyst's checklist for beginning to evaluate the importance of resources in the industrial location decision.

Reason 4: Transportation

Generally, the geography of the market that the manufacturing firm serves dictates the requirements for available transportation facilities. To be competitive globally, there must be an international deep-water port. To be competitive nationally, there must be national rail lines with spurs to the prospective industrial sites. To be competitive regionally, there must be interstate highway access to the proposed industrial sites. If management frequently flies, or if the finished good is transported by air, then the firm will likely be biased toward locating in a city where there is an airline hub (see table 6.6).

Reason 5: Government

National governments can bias firms to locate manufacturing facilities in their territory if they have a large market and if they can impose high tariff barriers. Henry Ford located Ford Motor Company in Detroit because it was only there, near his hometown, that he could obtain the necessary venture capital. Detroit was a fortunate location because across the Detroit River was the British Empire. An automobile assembled

Table 6.6. Illustrative Checklist for importance of transportation and conformity to the local market situation

Importance to the firm	Description of the local market area (situation)
Does the manufacturer sell to a global market?	Does the locality have a deep-water port?
Does the manufacturer sell to a national market?	Does the locality have quality freight rail service?
Does the manufacturer sell to a regional market?	Is the locality on an interstate highway?
What proportion of total costs is transportation of raw materials and finished product?	If the manufacturing firm locates at the site, will the total cost of transportation be the same or lower than at other competitive sites?

in the British Empire could be sold duty-free in any other member country. So the U.S. auto industry built assembly plants in Ontario and later in cheap-labor Quebec. With the demise of the British Empire, and the benefits associated with membership, there needed to be new reasons for the U.S. auto manufacturers to keep their investment in Canada. The Canadian federal government then negotiated with the U.S. auto manufacturers for a free trade agreement on automobiles. Their agreement was then presented to the U.S. Congress, which passed the initiative into law.

State and local governments have less ability than the national government to influence industrial location. Since a benefit of U.S. statehood is free and unfettered trade with other states, then tariff and tax barriers cannot restrain trade and therefore cannot be an incentive to locate a manufacturing facility.

The opposite of a tariff and tax is an incentive (table 6.7). Most economists and geographers do not support incentives because if they were effective, they would then bias the economy toward an inefficient spatial organization, thereby lowering the productivity and raising the cost of that productivity. Most evidence is that incentives do not in reality bias manufacturing from one location to another. Very large incentives would be required to significantly affect the cost of production of a large firm; local governments simply do not have the financial resources to significantly affect the cost of production. Still, the representative of a manufacturing firm may use his or her bargaining position to extract some of the financial gains that will result from the

Table 6.7. Illustrative checklist for importance of government and conformity to the local market situation

Importance to the firm	Description of the local market area (situation)
Does the firm have a low fixed cost for manufacturing facilities and could therefore be considered unrestricted in its location?	Governmental incentives can influence the location decision of an unrestricted firm, and when the incentives end, the firm departs for another location.
Does the firm consider incentives to be an important signal as to how it will be treated and received by the community at large?	Does the locality have the legal base for offering a sufficient incentive package that will overcome the firm's anxiety about the new location?

Table 6.8. Illustrative checklist of important amenities

- Education from kindergarten through advanced graduate level
- Low crime
- Parks and recreation
- Architecturally interesting urban-built environment
- Nearby and plentiful natural environment, separate and distinct from the urban environment
- Urban planning that ensures interesting and attractive architecture, availability of supporting infrastructure, land uses spatially distributed to minimize conflicting negative externalities
- Affordable housing
- Good access to quality shopping and to other desired destinations
- Public transportation

multiplier effect (see chapter 2). However, the success of such negotiations can make for a community that has fewer of the amenities required to attract labor.

Reason 6: Amenities

There has been discussion of the role of amenities in most of the previous five reasons for industrial location. Today it is impossible to separate variation in manufacturing costs, particularly labor, from amenities. What is considered an amenity depends on the culture of the household, household income, and even location. A short list of amenities is presented in table 6.8,[11] The list is presented in no particular order that would necessarily be agreed upon by all communities. A locality may have a goal to excel in each of these amenities, and others not included in table 6.8; but, it may not be feasible to do so.

Reason 7: Technology, Scope, and Scale

Closely affiliated with agglomeration is the technology, scope, and scale of the locality. Already mentioned was the importance of a locality's scope of employment opportunities for dual-income households. This category also serves to reference the culture of the locality, the innovativeness of the local population, the willingness of the local population to adopt new ideas and understand new technologies, and their capacity to be comfortable working with leading edge technology. Industry of the twenty-first century does not seek out local cultures that are resistant to change and resistant to new ideas. One example is the Internet. Internet access has become a fundamental component of infrastructure. Internet infrastructure has become a prerequisite for many types of industry in their location decisions.

Putting It All Together

All the above components are programmed into geographic information systems software. This information along with an appropriate database, is used by the firm and its advisors in location analysis to select the site and situation that best conforms to the objectives of the firm. The analyst thus must consider the criteria being used by

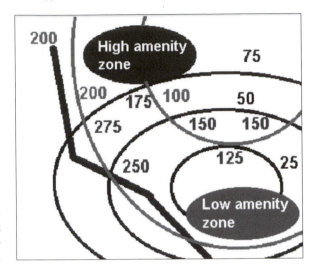

Figure 6.17 Hypothetical GIS-based analysis showing cost isolines and qualitative information.

manufacturing firms to select locations and be able to evaluate the prospects for any particular site and situation in how it conforms to the objectives of the industry.

Figure 6.17 is a hypothetical GIS analysis for a firm. Isolines, or contour lines, might represent the total cost of manufacturing at a location, including the cost of assembling inputs and distributing the finished product to market. Included with the cost calculation would be costs for power, labor, land, and so on. Such isolines are then overlaid on a map depicting various qualitative features, such as zones of high amenities, zones of quality schools, and so on. The decision makers then combine the qualitative and quantitative analysis to choose a site and situation (see, e.g., Warden 1993).

Concluding Remarks

The market analyst provides advice on the market for commercial space, including office and industrial properties. The analyst is responsible for the creation and assembly of information on market conditions for office and industrial space at a site and in the surrounding region. The advice is used to improve judgment on how much to charge for leased space, how much to sell the property for, or how much the property can be refinanced for, if commercial space should be purchased or rented, and at what price.

As with other real estate products, the analyst proceeds by defining a market area. In general, the analyst evaluates the market from the perspective of the large macro scale, and then down to the local or micro scale. The analyst performs the supply and demand analysis specific to the market area and finally assembles the information into a report for the client decision maker.

7

Retail

Retail real estate analysis is the most well developed, complex and techno-logical of any of the other real estate product categories.[1] This chapter begins with a review of the background literature. Real estate market analysis has been performed for retail longer than for any other category. Unlike medicine, in which some practices, such as bleeding a patient, have become obsolete, no major method that has been widely adopted by the industry during the past 100 years has subsequently gone out of use. Instead, these methods have been modified and incorporated into contemporary analysis and technology.

This chapter next proceeds to the macro level with a presentation of how a real estate location strategy is developed for a large, multibranch retail chain. Afterward, a brief discussion of the micro-level strategy is presented following the four steps introduced in the preceding chapters. The methods, technology, and analysis of the four steps for retail real estate have already been presented in chapter 4. Because the retail side has been such a pioneer, methods developed for retail ultimately find use, with some modification, in real estate market analysis of all the other real estate product categories; hence, the four steps were presented as part of the general methods of chapter 4.

Background

It has always been the case that certain locations for retail activities offer distinct advantages over other locations. But knowing which locations were best has not always been the complex task it is today. Until the late nineteenth century, the retail location decision was quite simple: always locate at the downtown commercial node.

Changes in transportation technology increased the geographic range of the individual household. Each successive transportation era brought increasing geodemographic complexity to the city. The eras of transportation can be broken down by transportation mode:

- First, households relied primarily on walking from home to work to shopping.
- Second, for some of the larger cities, the trolley car and other innovations in public transit brought an increased geographic reach of the average household.
- Third, the personal automobile allowed many households to move beyond the limited corridors of public transit.
- Fourth, beginning in the mid-1950s and continuing into the twenty-first century, the freeway has given the average household the geographic choice of hundreds of square miles.

Each successive transportation era has brought greater locational choice to the household, which has resulted in cities becoming more geographically complex. The increasing geographic complexity has been further compounded by a wider array of taste preferences and lifestyles within demographic categories.

In response to the increasing complexity of the city, retailing has adopted a "war room" strategy. The strategy is steeped in the latest computational and geographic technologies and locational analysis allowing retailers to track and hone in on their target demographic population. They keep track as new neighborhoods emerge and project the change in demographic composition in older neighborhoods. This information is used to improve judgmental decisions in corporate real estate.

A discussion of the procedures used in choosing a situation for a retail store was presented in chapter 4 as examples of general locational methodologies. The procedures include general hedonic modeling, gravity and spatial interaction models, customer spotting, the analog method, and the derivation of customer density surfaces and trade area boundaries.

After a situation has been chosen, the retail store has to choose between two fundamental options: to be a free-standing entity or to be part of an agglomeration of other stores in a retail center. The choice made depends on cost and expected benefits from each site. Retail centers, to compete for the retail stores, have diversified their offerings to various architectures with different value platforms. The value platform is the full experience the consumer receives from the purchase of the good or service.

Retail centers once could be classified as being neighborhood, community, regional, and super-regional centers. These original four categories date to the 1950s. Now the categories have grown to eight recognized shopping center types (see box 7.1). Diversification does not stop with these eight types. The city is continually changing, and in response, new forms of real estate projects are innovated; some are eventually adopted, and some are discarded. The recent success of the lifestyle center may result in that type of retail center becoming the ninth recognized category (see box 7.2).

Each type of retail center has a distinct function, trade area, architectural footprint, and tenant mix. The descriptions included in box 7.2 and table 7.1, such as size, number of anchors, and geographic range of trade area, are typical of each type of retail center and are not meant to exactly specify what is required for every center.

Box 7.1.   Categorization of retail centers

Shopping center: A group of retail and other commercial establishments that is planned, developed, owned, and managed as a single property. On-site parking is provided. The center's size and orientation are generally determined by the market characteristics of the trade area served by the center. The two main configurations of shopping centers are malls and open-air strip centers.

Basic Shopping Center Configurations

Mall: Malls typically are enclosed, with a climate-controlled walkway between two facing strips of stores. The term represents the most common design mode for regional and super-regional centers and has become an informal term for these types of centers.
Strip center: A strip center is an attached row of stores or service outlets managed as a coherent retail entity, with on-site parking usually located in front of the stores. Open canopies may connect the storefronts, but a strip center does not have enclosed walkways linking the stores. A strip center may be configured in a straight line or have an L or U shape.

Shopping Center Types

Neighborhood center: This center is designed to provide convenience shopping for the day-to-day needs of consumers in the immediate neighborhood. A supermarket anchors roughly half of these centers, and about a third have a drugstore anchor. Stores offering drugs, sundries, snacks, and personal services support these anchors. A neighborhood center is usually configured as a straight-line strip with no enclosed walkway or mall area, although a canopy may connect the storefronts.
Community center: A community center offers a wider range of retail goods than the neighborhood center does. Among the more common anchors are supermarkets, super drugstores, and discount department stores. Community center tenants sometimes contain off-price retailers selling such items as apparel, home improvement/furnishings, toys, electronics or sporting goods. The center is usually configured as a strip, in a straight line, or L or U shape. Of the eight center types, community centers encompass the widest range of formats. Some centers that are anchored by a large discount department store refer to themselves as discount centers. Others with a high percentage of square footage allocated to off-price retailers can be called off-price centers.
Regional center: This center type provides general merchandise, especially apparel, and a great scope of services. Its main attractions are its anchors: traditional, mass merchant, or discount department stores or fashion specialty stores. A typical regional center is usually enclosed with an inward orientation of the stores connected by a common walkway, and parking surrounds the outside perimeter.
Super-regional center: Similar to a regional center, but because of its larger size, a super-regional center has more anchors, a wider selection of merchandise, and draws from a larger population base. As with regional centers, the typical configuration is as an enclosed mall, frequently with multiple levels.

(continued)

Box 7.1. *(continued)*

Fashion/specialty center: A center composed mainly of upscale apparel shops, boutiques and craft shops carrying selected fashion or unique merchandise of high quality and price. These centers need not be anchored, although sometimes restaurants or entertainment can provide the draw of anchors. The physical design of the center is sophisticated, emphasizing a rich decor and high-quality landscaping. These centers usually are found in trade areas having high-income levels.

Power center: A center dominated by several large anchors, including discount department stores, off-price stores, warehouse clubs, or stores that offer great selection in a particular merchandise category at low prices. The center typically consists of several free-standing (unconnected) anchors and only a few small, specialty tenants.

Theme/festival center: These centers typically employ a unifying theme that is carried out by the individual shops in their architectural design and, to an extent, in their merchandise. The biggest appeal of these centers is to tourists; restaurants and entertainment facilities can anchor them. These centers, generally located in urban areas, tend to be adapted from older, sometimes historic, buildings, and can be part of mixed-use projects.

Outlet center: Usually located in rural or occasionally in tourist locations, outlet centers consist mostly of manufacturers' outlet stores selling their own brands at a discount. These centers are typically not anchored. A strip configuration is most common; however, some are enclosed malls, and others can be arranged in a village cluster.

From International Council of Shopping Centers (ICSC), *www.icsc.org*, used with permission.

A retail center is generally defined by its tenant mix, which gives rise to the types of goods and services sold and the size of the center, as described in boxes 7.1 and 7.2 and tables 7.1 and 7.2. A particular center may not fit within any of the general classifications. A hybrid center may combine characteristics of two or more classifications, or a highly unusual concept may have been used in the center's development. Other types of centers are not classified separately, but are a part of the industry.

At one end of the size spectrum of centers not classified separately, is the *convenience center*, which is generally small and contains tenants that offer a narrow mix of goods and services; the geographic range for their trade area is quite small. A convenience center anchor store might be a mini-mart convenience store. At the other end of the size spectrum are super off-price malls that consist of factory outlet stores, department store close-out outlets, and category killers in an enclosed mega-mall; these retail centers may have as much as 2 million square feet in an enclosed complex. The trend toward differentiation and narrowing offerings down to finely honed geodemographic submarkets result in a plethora of diversified retail centers, including home improvement centers, car care centers, recreational vehicle centers, and so on.

Not all sites and situations offer the same economic opportunities to a retailer, and not all design concepts will be successful. However, there may be a location where a

Box 7.2. The lifestyle center as a new shopping center category

The lifestyle retail center combines the convenience of strip center shopping with well-known specialty tenants found in regional malls. The lifestyle center is designed to appeal to highbrow customers who either lack access to or do not enjoy the shopping mall experience. The lifestyle retail center, lacking a traditional anchor, offers customers greater convenience and an improved shopping experience. The lifestyle retail center offers the tenant lower operating expenses and an environment that increases their customer patronage. Often lifestyle retail centers are combined with entertainment projects such as multiscreen movie theaters.

The lifestyle retail center is generally situated in the core of growing, high-income suburban areas. The lifestyle retail center offers a locational situation providing access to the tenants' target demographic populations, a synergy that arises from the tenant mix and from ambience created by the architecture of the lifestyle retail center and lower occupancy costs. Lifestyle centers are thought to work best in two scenarios: in mid-size markets where they function as the area's fashion center, and in large markets where they can fill a void in the absence of a regional mall.

Higher occupancy costs arise from payments for the benefits of being in the same retail center as a high-order anchor store. However, for some target demographic populations, the anchor store is not an attraction. The tenants of lifestyle retail centers then may avoid the high costs associated with anchored malls and even benefit from greater store patronage, as their customers are not avoiding them because of what to them is a negative shopping mall experience.

Evidence for higher revenues is that average sales per square foot at lifestyle centers are double that of regional malls. The average amount spent per visit at lifestyle centers is $107, compared with $70 at malls, and 45 percent of lifestyle shoppers rarely shop at regional malls.

Early lifestyle retail centers suffered because architects did not fully understand how people would interact with the retail concept. The wrong architecture can lead to a financial debacle. Because of the risk involved with new planned urban concepts like lifestyle retail centers, financing can be difficult. New concepts like this provide a real challenge to the analyst because there is not sufficient inventory at the start to execute an analoglike procedure (see chapter 4).

Examples of lifestyle centers include:
- One Pacific Place, Omaha, Nebraska, 90,000 square feet
- Town Center Plaza, Kansas City, Missouri, 700,000 square feet
- The Avenue of East Cobb, East Cobb, Atlanta, Georgia, 225,000 square feet
- The Avenue of the Peninsula, Rolling Hills Estates, California, 380,000 square feet

Mall stores located in lifestyle centers include Ann Taylor, Williams Sonoma, Talbots, Banana Republic, Nicole Miller, Eddie Bauer, The Sharper image, Pottery Barn, Liz Claiborne, The Gap, Restoration Hardware, Foot Locker, Bath & Body Works, Borders Books & Music, Abercrombie & Fitch, Nine West, and Smith & Hawkin.

Based on Kevin Kenyon. 1998. Will lifestyle name put wind in anchorless centers' sails? *Shopping Centers Today*, June. Available at *www.icsc.org*. Used with permission of ICSC.

Table 7.1. Typical acreage of shopping centers by concept category

Concept	Acreage
Convenience	3–15
General merchandise; convenience	10–40
General merchandise; fashion mall (typically enclosed)	40–100
Similar to regional center but has more variety and assortment	60–120
Higher end, fashion oriented	5–25
Category-dominant anchors; few small tenants	25–80
Leisure; tourist oriented; retail and service	5–20
Manufacturers' outlet stores	10–50

From International Council of Shopping Centers (ICSC), *www.icsc.org*, used with permission.

particular design concept will thrive, and other locations where the same retail concept will result in a financial failure. The real estate market analyst has the challenge of assembling the appropriate data and executing the appropriate analysis to improve judgmental decisions on a wide array of characteristics related to a project, including:

- Identifying the sites and situations that offer an advantage and discarding from further consideration those locations that are at a competitive disadvantage
- Evaluating the sequence of market penetration, proceeding from the macro scale, identifying which regions the firm should enter in which order, down to the micro scale, identifying local sites and situations
- Evaluating the performance of existing retail projects; identifying which should be targeted for expansion and which should be targeted for closing
- Evaluating the feasibility of a type of retail project at a particular site
- Determining the size of a proposed development
- Identifying market niches that are unfilled
- Recommending tenant mix
- Recommending timing of development
- Recommending defensive locational strategy.

The demand for the real estate market analyst is a direct result of three factors: the increasing complexity of the city, the importance of advice given regarding the above

Table 7.2. Typical anchors by shopping center category

No. of anchors	Classification	General description and frequent tenant mix	Typical primary trade area (miles)
1 or more	Community center	Discount department store; supermarket; drug; home improvement; large specialty/discount apparel	3–6
2 or more	Super-regional center	Full-line department store; junior department store; mass merchant; fashion apparel	5–25
N/A	Power center	Category killer; home improvement; discount department store; warehouse club; off-price	5–10
N/A	Outlet center	Manufacturers' outlet stores	25–75

From International Council of Shopping Centers (ICSC), *www.icsc.org*, used with permission.

list, and the fact that the analyst has available the new geographic technology to get the job done in an accurate and highly productive manner. Therefore, the real estate market analyst must have a specialized knowledge base and the technical skills necessary to unravel the geographic complexity of the city in a manner that is relevant to the decision maker.

There is a two-way interaction between the analyst and the geography of the urban-built environment. On the one hand, the analyst makes recommendations on projects based on the present and projected geodemographics. On the other hand, it is the judgmental decisions based on such analysis that create the geography of the city. The analyst therefore responds to, and is a cause of, the geography of the urban-built environment. It is thus necessary to understand how judgmental decisions on development were made in the past to understand the existing urban-built environment, including its geodemographics, development trends, and the potential for future infill development.

The Antecedents

Figure 7.1 and table 7.3 are a graphical and descriptive timeline for business geographic market analysis.[2] The stages begin with the methods of the nineteenth century and proceed through the high-tech era of the twenty-first century. Although each stage began at a different time, the general reasoning of each stage is still valid. In combination, the methods of all stages provide the basis for location analysis as practiced today.

The first stage begins with the business geographers application of pedestrian trial-and-error methods commonly used in nineteenth and early twentieth centuries. The second stage is the era of cartographic representations including the various contributions of geographer William Applebaum (1965a, 1965b, 1966, 1968; see also Applebaum and Cohen, 1960) that have already been explained in chapter 4. The third stage includes more sophisticated quantitative analysis, including gravity and spatial interaction models, also explained in chapter 4. The fourth stage is an integration of quantitative and qualitative reasoning. The fifth stage, beginning in the 1970s, is when

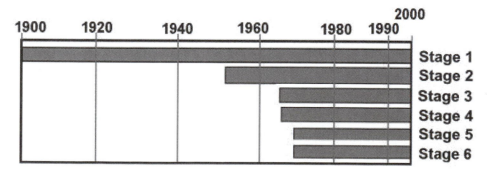

Figure 7.1 Time line for business geographic market analysis of retail location decisions (after Thrall and del Valle 1996a, 1997b).

Table 7.3. Timeline of location-based market analysis

Stage	Description
1. Pedestrian and sign-posting era	The city center offered the best location for marketing high-order retail goods. Other sites may have been chosen, but on a trial-and-error basis. Some parallels to this approach exist today; for example, some national retail chains consider only a location near the regional mall adjacent to the freeway exit (Thrall et al. 1996).
2. Mapping and cartography	The cumulative body of knowledge about locational analysis goes back at least to the 1920s and the use of gravity models back to the 1850s. Early applied business geography originates with contract work by retail chains to determine the relative value of a site, compared with the relative advantages of other sites. Most important in the development of this stage was the work of Applebaum. The methodologies used in the era before contemporary geographic technology were mainly subjective in nature (Goldstucker et al. 1978).
3. Sales potential predictive models	Began in the mid-1960s and contributed a number of methodological advancements, particularly in using and manipulating statistical data to arrive at sales potentials for a total complex of stores in a shopping center (Craig et al. 1984). Most important in the development of this stage of marketing geography were the contributions of Huff (1959, 1963) to the advancement of gravity and spatial interaction models. For discussion, see Haynes and Fotheringham (1984).
4. The rise of the holistic view	Began around 1966 and saw the development of more sophisticated models of retail potential that considered the influence of store size, distance, image factors, merchandise quality, and services of all competitors within the same trade area. During this stage, academic research largely had the objective of measuring retail image and the dimensions that underlie these perceptions; seldom were these considerations dealt with in the professional and academic locational analysis literature prior to the 1960s (Goldstucker et al. 1978). This stage differs from the third stage by the inclusion of nongeographic factors as having equal or even greater importance than location.
5. GIS as an exotic technology	The 1970s and 1980s were the beginnings of the use of geographic information systems in retail location analysis. However, the software in this era was very complex, often having been adopted from natural resource applications, or written from scratch by programmers at the retail firm. Data were either very expensive or not available. GIS had a deserved reputation in this era of being an exotic technology that was very expensive and difficult to use. The GIS of this era was dependent on mainframe computers or large workstations.
6. GIS as common desktop computer technology, integration of geographic analysis into routine strategic management	Noted for significant advances in desktop computer hardware, including CD-ROMs for the distribution of data: the widespread use of Microsoft Windows operating system that allowed for the standardization of the software user interface: declining computer software programming cost; and the emergence of a business geography technology industry offering a wide array of low-cost data and specialized, highly productive software applications. The future of this stage will include the integration of GIS and management practices, a greater sophistication of the modeling behind the analysis, automation, improved qualitative analysis, and real time information delivered via the Internet.

early geographic technology began to be applied to improve the business decision. However, although the technology existed, its high cost and great complexity precluded its adoption by all but a few innovative firms and academics. The sixth stage, beginning in 1990, is beginning of the contemporary desktop GIS era; the era is accompanied by low-cost, high-quality software and a support industry particularly for data that revolutionized the business geographers' practice of retail real estate market analysis. Each stage is explained in greater detail below.

Stage 1: Pedestrian Era

Dating back millennia, the many cultures of the earth have taken interest in, developed procedures, and applied those procedures to choosing locations, site, and situation to carry out human activity, including retail enterprise. Retailers have occasionally relied on a trial and-error approach: pick a location at random and see if the site meets one's objectives. If it does not meet objectives, then close it down and try another site. Those firms that follow the trial-and-error approach will last as long as their luck and their finances hold. The trial-and-error approach maximizes costs.[3]

The term *pedestrian* has been used to describe the first era of business geographers performing retail real estate market analysis, and comes from the reliance on pedestrian traffic volume count, which dominated the era. Because walking was the dominant mode of transportation, the business geographer analyst would count the number of persons who walked by a spot by the hour and by the day. The locations with the greatest pedestrian counts revealed the most favorable locations for the proposed activity.

The geographer William Applebaum (1966), whose methods were presented in chapter 4, considered the beginning of the discipline to date to late in the nineteenth century and early twentieth centuries. At that time, chain companies, primarily tobacco shops, began to conduct detailed surveys of pedestrian flow along streets to identify the most desirable sites (highest foot traffic) in the center of town (Davies 1977).

Trial-and-error and the pedestrian approaches lacked the ability to explain the cause for success or failure. Pedestrian approaches essentially apply what can be referred to as correlational *signposting*.[4] An example of correlational signposting would include the retail location strategy of a chain of dry cleaners. Dry-cleaner management knows that their success depends on being located in a neighborhood strip center (see box 7.1) adjacent to a grocery store. The grocery store in a strip center is the only signpost needed by management to indicate the likely success of a new project. The correlational signpost approach remains one of the most frequently used methods for selecting sites today.

Stage 2: Cartographic and Mapping Era

With the advent of the trolley car, other forms of public transportation, and then the affordable private automobile, the urban geographic landscape changed from earlier, simpler geographies. The geographic patterns in which people lived and shopped became increasingly complex (Epstein 1978). As cities spread outward, business geographers began to be increasingly relied upon to advise retailers where to locate. With the coming of the Great Depression in the 1930s, choosing the right location became inseparable with fiduciary responsibility. Whereas stage 1 can be characterized as having compact urban environments, stage 2 began the era of less compact, more dispersed urban environments.

With the rise of the suburbs, particularly in the post–World War II era, retail location decisions became much more difficult. No longer could the retailer simply locate a store at the signpost of the central business district. The methods that arose in stage 2 can be characterized as empirical descriptions of trade areas and market share that a store secured from its trading area (see, e.g., Applebaum 1968; Converse 1949; Douglas 1949a; Ferber 1958) These methods, still widely practiced today, depict the trade area cartographically. The cartographic representations were devoid of methodology for explaining and predicting consumer behavior. In many respects, cartographic representations of trade areas served to diminish the reliance the analyst may have had on a single signpost and introduced a methodology for evaluating many correlational signposts distributed about the trade area.

Cartographic representations remain the starting point for many real estate market analyses. The kernel method introduced in chapter 4 begins with a cartographic method, Applebaum's customer spotting procedure, and then proceeds to calculate a customer density, or revenue origin surface that can be used to project revenues based on that cartographic representation. The most significant contribution of the stage 2 era that has been passed down to today are the procedures discussed in chapter 4 and introduced by Applebaum (1966). Cloaked in new robes of technology, those methods still provide much of the basis for real estate market analysis used today. Contemporary variations on Applebaum's methods of customer spotting, trade area delineation, market penetration, and analog procedure include the use of geographic technology for mapping such data as ATM records, check records, credit card data, and point-of-sale data (usually ZIP code).[5] Knowing where their customers live, work, or otherwise originate from is central to real estate market analysis today. The analyst then applies this knowledge to make recommendations on where the resources of the firm can be most effectively allocated.

Using contemporary variations on the methods that arose during stage 2, analysts today are able to consider cannibalization that results from overlapping trade areas and missed opportunity that results from unserved trade areas. In the process of revealing the origins of customers, customer spotting also reveals the distance customers will travel to patronize a store. Cartographic representation of the origins of customers is a prerequisite for calculating the distance customers are willing to travel to a retail store, also known as the *range* (Berry and Garrison 1958; King 1984). The rate at which demand declines with distance to the retail center is known as the *distance decay* of store patronage. By applying these concepts, management's decision strategies may be made so that limited resources can be used to maximum potential.

Stage 3: Sales Potential and Predictive Models

In the 1940s, land economist and real estate professor and practitioner Homer Hoyt outlined the basic steps required for performing market analysis for shopping centers. At that time, shopping centers were a newly emerging phenomenon in the United States. Nevertheless, the steps Hoyt outlined remain part of the core of the business geographers' analysis of retail centers today. An expanded and updated list based on

---

Box 7.3.   Basic market analysis steps for shopping centers

- Analysis of the economic base of the metropolitan area, showing general characteristics of the market, including overall economic trends, employment trends, projections of economic activity, and growth patterns
- Delineations for primary, secondary, and peripheral trade areas and accessibility to them
- Population data for each trade area, including existing sizes, historic trends, and future projections
- Demographic data for each market segment targeted and information about the resident population in the trade areas including tourists, office workers, and convention and business travelers
- Population characteristics for each trade area including the number of households, families, and singles, lifestyles, age cohorts, historic trends, and future projections
- Income characteristics for each trade area including household, family, and per-capita totals and disposable income, purchasing power, and future projections (today, three year, five year and ten year)
- Patterns and trends of expenditures by type of goods and services in the trade areas
- Location, characteristics, and sales of competitive retail centers, by type of center, in the trade areas
- Availability and absorption of retail space and sales trends by retail categories in the trade areas
- Characteristics and status of proposed and planned retail developments in the trade areas
- Neighborhood and site characteristics if a specific site has already been chosen or comparisons of sites if multiple sites are under consideration
- Capture rates, productivity rates, and recommended characteristics/anchors/sizing of the center or centers depending on the scenarios being considered.

Based on Hoyt (1949, 1958) from Urban Land Institute (1999).

Hoyt's original steps, are presented in box 7.3. Each of the items listed in box 7.3 requires an accurate assessment and calculation of a geographic market area. Methods for evaluating each of the items have already been presented in preceding chapters of this book.

Continuing with the time line of figure 7.1 and table 7.3, Applebaum's customer-spotting method would be included in stage 3, as would be the beginnings of the popular application of gravity and spatial interaction models. While gravity and spatial interaction models were widely adopted during stage 3, their innovation in retail real estate market analysis began in the mid-nineteenth century, when geographers first applied Newtonian physics to measure social phenomena, including market areas.

Stage 4: The Rise of the Holistic View

There is much more to customer patronage than location. Of equal or perhaps even more importance are store image, managerial effects, number of checkouts, general service level and quality, and so forth. Simmons (1984) stated that if a store is achieving a low share of trade, it is vital to establish the image of the store compared with selected competition to find out why people shop at one store rather than another. This is an appropriate situation for a focus group, where a population of prospective customers are queried about their attitudes and shopping behavior.

Stage 5: GIS as an Exotic Technology

GIS (geographic information systems) began in the 1950s as an exotic technology whose development was stimulated by the Soviet launch of the Sputnik satellite. The United States followed shortly thereafter with their own satellites. But soon governmental agencies whose funding procured these satellites began to ask, what is their purpose? Retrospectively, the answers might seem obvious, but in the 1950s the future was not all that clear. Geography professor Evelyn Pruitt of the University of Louisiana was hired by the U.S. Navy Satellite Program to answer this fundamental question. Among her grand visions for satellite technology arose the concept and the term *remote sensing*; today every weather report routinely presents remote-sensing satellite images.

As early as the 1970s, some academics utilized the concepts of what would become GIS and applied those concepts to the human-built environment. For example, Thrall (1979) evaluated the quality of local property tax assessment and was the first to map the assessed value-to-market value ratio. Those neighborhoods that appeared as mountain tops were identified as being overassessed, while those appearing as the bottom of a valley were identified as being underassessed. However, in the 1970s, such analyses required several years to produce and therefore were not feasible to conduct on a routine basis, especially by commercial enterprise.

In the 1980s many counties and cities began to convert from printed maps for their legal property boundary records to maps in electronic digital format. About this time, a few notable corporations, most importantly McDonald's Restaurant Corporation, began to apply GIS and to integrate GIS into corporate location strategy. McDonalds hired a team of geographers and computer programmers to write their own proprietary GIS software. The many millions of dollars of expense was justified as they recognized that theirs was a location-dependent retail business, and being the first fast-food restaurant with a GIS war room gave them a substantial advantage over their competition.

Stage 5 of the 1980s is characterized as the time that some innovators recognized the potential that GIS would have to retail location and market analysis. However, few could adopt GIS because of the great complexity and high expense.

Stage 6: Desktop GIS and Business Geography Applications

While all the other stages have fuzzy beginning dates, this is not so for stage 6. Stage 6 began with the release of the 1990 U.S. Census TIGER/Line data. The TIGER/Line data consist of the entire street network of the United States and its territories in electronic digital format. Included with the digitized street network are attribute data for the names and addresses of the streets. The U.S. government placed the TIGER/ Line data into public domain.

Desktop computers of the early 1990s were undergoing dramatic increases in computational power and storage capability. Most important to geographic data storage and distribution was the CD-ROM. GIS software of the early 1990s, however, remained difficult for most users to master until the release of Microsoft Windows 95.

Microsoft Windows 95 and Apple Computer standardized the graphical user interface for the broad consumer and business markets. The most important decision a computer user makes is choosing which operating system to use. GIS software vendors effectively ignored Apple Computer, thereby making Microsoft Windows the standard interface. After the choice of operating systems, all software within the same category of purpose effectively looks the same—master one and you master them all. Prior to Microsoft Windows, just mastering a user interface of a typical workstation-based GIS software program was estimated to require three years of training. After Windows 95, GIS became a desktop management tool no more difficult to use than a spreadsheet or database management software programs. As an increasing number of users adopted GIS software, the threshold to support an industry of geographic technology and geographic data was created. Soon, a vast array of low-cost geodemographic data was distributed on CD-ROM. GIS has become a $50 billion per year industry.

The distinguishing characteristic of stage 5 is that the array of analysis that would be applied in real estate market analysis at the start of the twenty-first century, was identified; however, it was seldom actually applied in the business environment during stage 5 because of its great complexity, low productivity, and high expense. With the advent of stage 6, complexity, productivity, and expense were no longer an issue. The big issues of stage 6 became, how are you going to use the technology, how is the technology to be integrated into the managerial decision, and how is the technology going to benefit the business? (See Longley and Clarke 1995; and Castle 1998.) Now the impediment to adoption is lack of knowledge of geographic technology and geographic analysis by American industry, as well as by traditional disciplinary university faculty in business schools.

Macro-Level Corporate Location Strategy

In this section, the importance of a corporate location strategy is explained. From a top-down viewpoint, the location strategy evaluates the importance to a firm of which region to enter and in what sequence. This is differentiated from the micro level strategy of where within a market area a retail facility should be situated. For an international perspective on this see Sharkawy, Chen and Pretorius (1995).

Each retail chain targets its products toward certain types of households. A good location adds to the value of these products. It is one of the most important competitive advantages that companies can possess. In chapter 4 it was explained how real estate market analysts calculate the types of households likely to be attracted to a retail chain's products. Because consumers with targeted characteristics are not uniformly spatially distributed, the location of a retail chain affects the proximity between the targeted population and the site. Proximity affects demand for the goods and services that a retail chain offers. A good location, then, can help maximize a company's potential sales.

Retail chains must first determine who their target consumers are before expanding. Managers must be familiar with their company's *value platform*, which is the value of the range of products offered by the retail chain within the mind of the consumer. The value platform of a retailer arises out of the full experience the consumer receives from the purchase and the consumption of the good or service. Included in the value platform is the situation within which the retail facility is located. The value platform includes the service received from the sales personnel and the display of the goods at the retail store. The value platform also includes the image of the products sold as well as the benefits received by the consumer from the good itself. The value platform, then, includes the entire shopping experience. Retail marketing strategy has its foundations in the firm's value platform. The value platform specifies the manner in which the firm differentiates itself from its competitors in the minds of the consumers it intends to serve. An understanding of the value orientation of customers—the way in which they trade off shopping experience with price—is the basis on which firms make target market decisions (Ghosh and McLafferty 1987). Therefore, knowing the value platform and identifying the company's competitors and target market are fundamental when establishing a corporate marketing strategy.

A company's marketing strategy encompasses how it responds to product development, competition, pricing, and location policy. A company's location policy defines the trade areas in which it will participate. The resulting geographic trade area places a limit on the market potential (Ghosh and McLaffetty, 1987). Site analysts determine the value that a site adds to a retail chain (Thrall and del Valle, 1996a, b, c, 1997a, b). After identifying those characteristics of location that significantly influence retail sales potential, those location characteristics can be incorporated within a retail chain's location policy. Later in this chapter an explanation is presented for how location strategy is developed for a multibranch retail chain. When deciding on a location policy, managers must consider how prospective consumers will trade off distance with attraction. Their perception of the store, the price of the product or service, the shopping experience, and myriad psychological factors all influence a consumer's attraction to a retail outlet. The advantage that superior locations can bring to a retail chain can only be realized after the important nonlocational factors are controlled.

A well-designed and properly implemented location policy is a corporate asset. Although companies may make a profit without one, an appropriate location policy can increase profitability and competitive edge. As Davies and Rogers (1984, p. 164) wrote, "it is the sales above and beyond what the site can create by itself, so positioned in the market place, [that] actually produces the profitability of the retail unit." Retail chains with a substantial value platform may make a profit at any location.

Consider the implications for companies without a location policy. According to Davies and Rogers (1984), when a company's future is at stake, retail location research is so important and the price so small relative to the total investment that such research must be done. If a company had a monopoly, any site would suffice and likely produce a profit; however, if competitors have a sound location policy and a well-designed procedure for its implementation, companies without a location policy will be operating at a competitive disadvantage.

Ghosh and McLafferty (1987, p. 27) wrote that three levels exist at which competition occurs in the retail sector:

> [C]ompetition exists among groups of stores offering different types of products and services. An example is the competition among bowling alleys, game arcades, and amusement parks for consumers' expenditures on leisure activities. . . .
>
> competition exists among strategic groups offering similar types of merchandise. For example, off-price clothing outlets compete with both traditional department stores and upscale discount stores. . . .
>
> competition occurs among stores within a particular strategic group. These stores offer similar kinds of merchandise at similar levels of price and quality. Competition among these stores is usually intense because they aim at the same target group and each tries aggressively to gain market share. *The competitive advantage of firms offering similar merchandise and shopping experiences is largely determined by the price they charge and the location of their outlets.*

Changing prices is much easier than changing the location of a retail outlet. Therefore, investing in market area and location analysis has the potential to bring higher profit margins in the long run. It may make the difference between turning a profit, breaking even, or going broke. Also, because the costs invested into land and buildings (even when leased) are likely to be high, mistakes in location are expensive and difficult to overcome. Companies must often live with suboptimal sites (and therefore, suboptimized profits) simply because changing locations cannot be justified monetarily. Making the right choice in the first place can be extremely advantageous to a company.

A Retail Chain Expansion Strategy

This section introduces an approach to the development of a multibranch retail chain expansion strategy. The retail real estate market analyst is responsible for analysis at the macro scale down to the micro scale. At the macro scale, the type of questions the analyst needs to address include which regions should receive the retail sites first and what should be the level of market saturation before another region is opened for penetration. At the micro scale, the type of questions the analyst needs to address include where within the region the first and each successive retail branch should be located; how much cannibalization will occur to existing branches' revenues if an additional branch is located at a particular site; and how are existing branches performing based on various criteria.

The seven-step multibranch retail chain expansion strategy of Thrall et al. (1997, 1998a, b, c, d, e) and Thrall (1998b) is presented below using data provided by the Darden (Red Lobster) Restaurant chain. Their methodology is most appropriately used by retail chains that

- Have multiple outlets
- Are planning to add more outlets;
- Target specific demographics
- Have reasonably accurate measures of their products' reach (i.e., how far or how long consumers are willing to travel to obtain the company's goods or services)
- Have maintained a database of annual performance measures for each outlet
- Have analysts on staff, or have some affiliation with a consulting firm that can perform the required analysis.

Seven steps comprise the multibranch retail chain expansion strategy:

- Identify the broad region for store assessment
- Perform a geographical inventory of competitors
- Assess the relative performance of competitive retail units
- Identify situation targets—namely, the geodemographics and environmental character-istics (both natural and built) that surround the site
- Assess market penetration
- Identify geographic markets and submarkets for expansion
- Make judgment.

The Thrall et al. methodology relies heavily on GIS technology. GIS makes the process of spatial analysis more efficient and accurate, allowing companies to identify consumers' locations, their competitors, and which retail sites have the greatest profit potential. When GIS is used to analyze a retail site, companies can attain a competitive advantage over others that have yet to adopt appropriate GIS technology and geographic analysis.

Step 1: Identify Area for Store Assessment (Expansion Strategy)

How should a retail chain develop its network of outlets? The first step deals with the geographic scale that the retail chain will attempt to cover in a fixed period of time. This period of time is usually defined within a company's business plan. The company stipulates in its business plan funds that can be allocated to capital expansion. This amount translates into a specific number of outlets that can be added. The question then is, where to put them? Should the chain attempt to gain a significant market presence over the entire nation? Financing and other resource limitations, including personnel, may prohibit a retail chain from following this comprehensive approach.

Should the retail chain proceed from region to region only after each region is saturated? If the company follows a region-by-region expansion approach, what should be the geographic scale of the region: the entire southeast or only the north-central portion of a state? Financing and other resource limitations are the initial constraints. The parent company may not, by understanding its resources and limits, be able to place outlets into market areas beyond its ability to finance, service, and manage those branches. Therefore, new companies with limited resources may be constrained to focus on a comparatively small geographic area. When resources allow for expansion, companies may expand into nearby market areas. The geographic expansion strategy is not a chicken-and-egg problem. Financing and resources come first; afterward, the feasible strategy for geographic expansion is identified.

Within the active geographic market area, should the company proceed from the largest markets and move down the hierarchy of cities (King 1984)? Or should it proceed from the smallest market area that can support the outlet and move up the hierarchy of cities (see Morrill et al. 1988)? Multibranch retail chains have successfully followed both approaches. Wal-Mart (Bentonville, Arkansas) has been successful partially because of its decision to expand first into smaller-sized market areas that had been overlooked and avoided by competitors, and then into larger market areas. Pavilions, an upscale division of Vons Grocery Stores (Arcadia, California), has instead expanded first into the wealthier and larger market centers, but have remained geographically focused on the West Coast. Starbucks, the world's largest purveyor of specialty coffee, expanded first near its home in the U.S. Pacific Northwest, then leaped to the higher order U.S. urban centers including Chicago, New York, and Los Angeles. After achieving a national image, Starbucks then worked its way down the hierarchy of U.S. cities. Starbucks is now down to urban centers in the United States with a population size of 100,000. Starbucks is also beginning its expansion into the European, Asian and Latin American markets.[6]

The expansion strategy followed depends especially on the analyst's assessment of existing market penetration by competing firms and the density and distribution of the population that the retail product will be marketed to. In the evaluation of the expansion strategy, the analyst may perform an assessment of those characteristics referenced in box 7.3, but for the smaller, lower order cities in the market to be penetrated, and then for the larger, higher order cities in the market to be first penetrated. The two scenarios would then be compared and contrasted as to which is more likely to result in higher profits for the retail chain.

Step 2: Perform a Geographical Inventory

A geographic inventory is a listing of facilities by submarket. In table 7.4, the submarkets are U.S. Census MSAs (metropolitan statistical areas), and the facilities are branches for Red Lobster restaurants in 1990. The geographic inventory can bring to the attention of the analyst gaps that may exist in some submarkets, or too great an exposure in other submarkets. Red Lobster Restaurants had not begun to penetrate the Salt Lake City MSA in 1990. Yet Salt Lake City was the 37th ranked MSA in the United States, and Red Lobster restaurants had already penetrated MSAs with fewer people. The geographic inventory brings this to the attention of the analyst in a qualitative manner, but does not explain why. Stiff competition of long-established seafood restaurants may have justified holding off penetration in the Boston MSA submarket.

In addition to presenting the geographic inventory as a table, the facilities can be mapped by submarket or by some geographic aggregation of submarkets. The geographic visualization via GIS of the outlets can affect a manager's decision-making processes by allowing the locations of an entire set of retail outlets to be displayed and compared to other simple thematic maps. The result may be displayed as a thematic dot map showing the density of retail outlets by submarket, such as a state or county, or as a traditional pin map where each pin represents a retail outlet and a particular color represents a specific brand. Either kind of map can reveal trade area presence. Figure 7.2 is a map of Red Lobster restaurants by state in 1990, which is an aggregation of Red Lobster restaurants by MSA within a state. Figure 7.3 is the projected per-capita food expenditures outside the home for the year 2002.

It would be erroneous to evaluate the Red Lobster geographic inventory of figure 7.2 by comparing it only with the projected 2002 per capita out-of-home food expenditures of figure 7.3 because many characteristics are used in the site and situation evaluation of a food retail outlet. However, the comparison may signal the analyst that additional consideration might be given to states such as Colorado, with comparatively few restaurants, but with a high propensity to eat food outside the home.

Geographic inventories are not limited to the brand of the client or employer. Data on competitive brands are readily available. Any market is limited in ability to absorb

Table 7.4. Sample geographic inventory of Red Lobster Restaurant locations in 1990 in U.S. Metropolitan Statistical Areas (MSA)

MSA	No. of rank units per MSA	MSA name
1	20	New York-northern New Jersey-Long Island, NY-NJ-CT
2	24	Los Angeles-Anaheim-Riverside, CA
3	23	Chicago-Gary-Lake County, IL-IN-WI
4	6	San Francisco-Oakland-San Jose, CA
5	8	Philadelphia-Wilmington-Trenton, PA-NJ-DE-MD
6	15	Detroit-Ann Arbor, MI
7	8	Washington, DC-MD-VA
8	13	Dallas-FT. Worth, TX
9	0	Boston-Lawrence-Salem-Lowell-Brockton, MA
25	6	Kansas City, MO-KS
26	2	Sacramento, CA
27	2	Portland-Vancouver, OR-WA
28	4	Norfolk-Virginia Beach-Newport News, VA
29	5	Columbus, OH
36	14	Orlando, FL
37	0	Salt Lake City-Ogden, UT
47	5	Jacksonville, FL
48	1	Albany-Schenectady-Troy, NY
49	2	Richmond-Petersburg, VA
50	4	West Palm Beach-Boca Raton-Delray Beach, FL
51	0	Honolulu, HI
52	0	New Haven-Waterbury-Meriden, CT
144	0	Duluth, MN-WI
145	1	Huntsville, AL
146	1	Tallahassee, FL
147	0	Anchorage, AK
148	1	Roanoke, VA
149	2	Kalamazoo, MI
150	1	Lubbock, TX
151	1	Hickory, NC
152	1	Lincoln, NE
153	1	Bradenton, FL
154	1	Lafayette, LA
155	1	Boise City, ID
156	1	Gainesville, FL
265	0	Iowa City, IA
266	0	Elmira, NY
267	1	Sherman-Denison, TX
268	1	Owensboro, KY
269	0	Dubuque, IA
280	0	Casper, WY
281	0	Enid, OK

product; the level of absorption might be taken by one's own brand or the competition. Thus, the geographic inventory of the competition must be tracked as well. Retail brand inventory is available through inexpensive sources such as yellow pages on CD-ROM products. There are many such products available at office supply stores.

They allow for queries by brand name or standardized industrial code (SIC); some

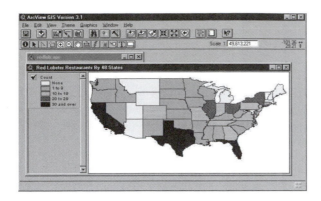

Figure 7.2 Geographic inventory of Red Lobster restaurants by state (Thrall et al. 1998a).

even allow mapping the results of the query. These data products allow exportation of the database query to common formats that may be then imported into GIS software. The data may then be geocoded, mapped, and analyzed.[7] If the retail chain owns the building, then F. W. Dodge data, as shown in chapter 4, may be used to select by brand name and output pipeline data for that brand by submarket.

Step 3: Assess the Relative Performance of Retail Units

Visualizing geographic corporate data can begin the process of evaluating geographic expansion or contraction strategies. Slightly more complex thematic maps that scale the radius of the dots in proportion to the revenue or profit of each outlet can further aid a manager's judgment. Additional descriptive information can be shown as pie charts.

Data on individual branch performance within a multibranch retail chain normally comes from proprietary records kept at corporate headquarters on each branch. The data set of all branches may be geocoded using standard GIS procedures, and then the attributes can be mapped to allow geographic visualization. Data fields normally tracked by branch include branch revenues, number of customers, LSPs of customers, operating expenditures, rent, address, and geographic coordinates.[8]

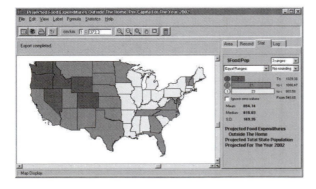

Figure 7.3 Projected per-capita 2002 food expenditures outside the home.

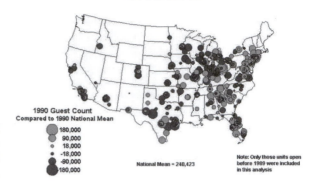

Figure 7.4 Guest counts in all Red Lobster restaurants compared to national mean in 1990 (Thrall et al. 1998).

After the proprietary or comparative data are acquired, the analyst when assessing the relative performance of branch facilities first will construct a map of guest counts in all branches. This was done using Red Lobster Restaurant data in figure 7.4. The larger the circle, the greater the deviation the individual restaurant's guest count is from the national mean guest count. The lighter shaded circles are restaurants that are above the national average. The darker shaded circles are restaurants that are below the national average. A map like that of figure 7.4 allows the analyst to view the system as a whole as opposed to looking only at individual branches. The ability to look at the system as a whole, with each element in the system showing its location and relative performance, is one of the greatest benefits to qualitative assessment obtained from the use of GIS. The map draws attention to individual branch relative performance, performance among branches on a regional level, clusters of branches with similar relative performance, and opportunities arising from gaps in the map.

Next, the analyst assesses the trend of performance of branch facilities. Performance trend is demonstrated in figure 7.5 with a map of change in guest counts by branch for all branches, again using data from Red Lobster restaurants. The larger the circle, the greater the change of the individual restaurant's guest counts. The lighter shaded circles are restaurants whose change is positive. The darker shaded circles are restaurants whose guest count is on the decline. Again, a map like that of figure 7.5 allows the analyst to view the system as a whole as opposed to looking only at individual branches. The map draws attention to trends in individual branch performance, performance trends among branches on a regional level, clusters of branches with similar performance trends, managerial changes that may need to be made, possible changes in regional taste preferences, and perhaps increasing competition on the regional level that may otherwise have been overlooked until later stages of the competitors' market penetration. It may also signal that some markets are oversaturated, while others may receive greater penetration.

Next, the business geographer analyst will combine the relative performance map of figure 7.4 and the trend map of 7.5 to construct a multitheme map like that in figure 7.6. Instead of shuffling back and forth between two sets of maps, qualitative

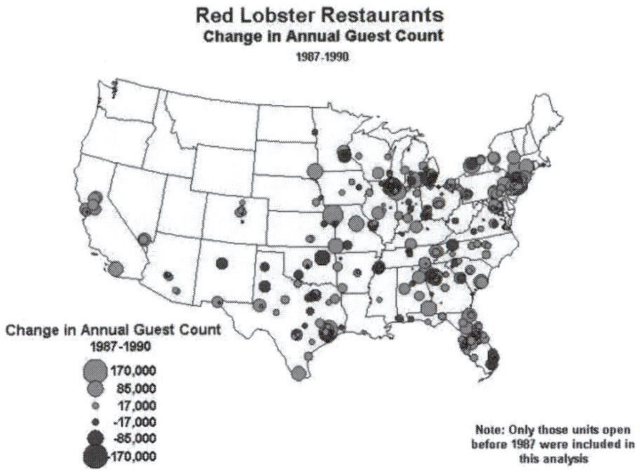

Red Lobster Restaurants
Change in Annual Guest Count
1987-1990

Change in Annual Guest Count
1987-1990

170,000
85,000
17,000
-17,000
-85,000
-170,000

Note: Only those units open
before 1987 were included in
this analysis

Figure 7.5 Change in guest
counts in all Red Lobster restaurants between 1987 and 1990
(Thrall et al. 1998b).

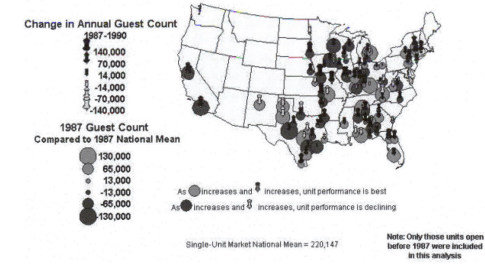

Red Lobster Restaurants
An Analysis of 1987 Guest Counts and
Change in the Annual Guest Counts From 1987-1990
Single-Unit Markets

Change in Annual Guest Count
1987-1990

140,000
70,000
14,000
-14,000
-70,000
-140,000

1987 Guest Count
Compared to 1987 National Mean

130,000
65,000
13,000
-13,000
-65,000
-130,000

As ⬤ increases and ⬆ increases, unit performance is best

As ⬤ increases and ⬇ increases, unit performance is declining

Single-Unit Market National Mean = 220,147

Note: Only those units open
before 1987 were included
in this analysis

Figure 7.6 Multitheme map for qualitative assessment of individual branch and regional trends
for single-branch markets, Red Lobster Restaurant data for 1987 and 1990 (Thrall et al. 1998b).

Table 7.5. A GIS categorical assessment of unit performance

	Map symbol type	
Map symbol type	Large dark-gray tack	Large light-gray tack
Large light-gray circle	(a) Sustained high performance	(c) Declining high performer
Large dark-gray circle	(b) Improving poor performer	(d) Sustained poor performance

Map symbol types refer to symbols used in figure 7.6.

assessment is enhanced with the GIS multiple-theme mapping capabilities. The two themes shown in the map within figure 7.6 combine performance within the target year of 1990 (figure 7.4) and performance relative to the four-year time horizon of 1987–1990 (figure 7.5). The interpretation by size and shade remains the same for figure 7.6 as that introduced in the two previous figures. However, instead of a circle being used as a symbol as was the case in Figure 7.5, an object shaped like a tack is used.

Figure 7.6 is a powerful tool for the analyst to make a qualitative assessment of individual branches and regional performance. The benefits to the analyst of being able to geographically visualize the relative and individual performance and performance trends of every branch in the system simultaneously cannot be overstated. The analyst is enabled to make powerful qualitative judgments about individual branches, regional clusters, and national trends.

In figure 7.6, branches that were indicated as performing at below the national average in 1990 (figure 7.4) did not necessarily show up as poor performers over a four-year time horizon. Note that figure 7.6 is a subset of all branches, as it shows information only for branches in single-unit markets. A computer graphics display or large color-print format would be required to show the detail for a multitheme map of a large organization like Red Lobster restaurants.

In figure 7.6, the higher performing the restaurant unit, the larger both the radius of the light gray circle and the size of the dark tack symbol. The lower performing the restaurant unit, the larger both the radius of the light gray circle and the size of the light gray tack symbol. Table 7.5 categorizes the possible qualitative assessment that the real estate market analyst may make based on the map in figure 7.6.

In figure 7.6, Red Lobster units in Florida have much higher than average 1987 guest counts as well as above national average increases in annual guest counts. In terms of table 7.5, the Florida market is revealed by the GIS analysis to be a strong category a-type. With such a concentration of category a-type units, the real estate market analyst might recommend further market expansion of branches within the Florida submarket.

In figure 7.4, Red Lobster units on the West Coast are far below the national average performance. And if a decision had been made in 1990 using that criterion alone, the analyst may have recommended closing those branches. Such a decision would have not necessarily been correct, as figure 7.6 and table 7.5 reveal that region to be category b-type, or improving poor performers. With category b-type units, management may be advised to consider a long-term assessment of the regional market. The West Coast market, while performing below the national average, also has units that are among the leaders in improving guest counts. Therefore, the GIS analysis indicates that the short-term poor performance might be overlooked in favor of long-term market development.

Northwestern Texas and New Mexico are regions revealed to be category c-type or declining high performers. Because of the geographic clustering of units there that are c-types, it is reasonable to develop initial hypotheses that some component within the region is responsible for the decline: perhaps poor quality regional management, a shift in regional taste preferences, the entrance of a successful competitor within the region, or oversaturation of the regional submarket. In a multiunit market, management may have caused the decline among formerly successful units by saturating the region with additional units; management may have believed that there was stronger potential for future growth and success than proved to be the case. So GIS allows the spatial association to assist in interpreting individual units, how multiple units perform as a result of their interaction within a market, and evaluation of the overall performance of a regional market.

Units in northern Illinois and north-central Texas stand out as candidates for assignment to category d-type. Category d-types are units that have had well below national average performance within a given year and declining guest counts. Reasons for the poor performance might include management, regional taste preferences, or competition. In addition, the units might have an unfavorable location, perhaps having succumbed to neighboring development that detracts from patronage.

The GIS analysis displayed in figures 7.4–7.6 allows the real estate market analyst to visualize their national market at one glance, comprehensively. The analyst can discern between individual outlets as to their performance following national standards and their trends. Potential trouble spots are readily identified. Places of particularly favorable regional opportunities are also identified.

Step 4: Identify Situation Targets

The *situation* of a retail outlet is the geographic context within which the site resides. It includes the demographics of the trade area, the inventory of nearby businesses (see Benjamin et al., 1998), the characteristics of the built environment (Thrall et al. 1995) such as historic urban core or exit on an Interstate, and other environmental phenomena that may affect its performance.

If the multibranch retail chain has an existing inventory of branches, then the analyst will solve for those situational characteristics. There may be several groupings of situational characteristics, such as characteristics of the subset of branches that are urban infill versus characteristics of the subset of branches that are interstate oriented.

If the multibranch retail chain does not have an existing inventory of branches, then the analyst will solve for those situational characteristics of a competing multibranch retail chain. The analyst will be effectively reverse engineering the geodemographic characteristics sought after by the competition. The analyst may discover that several location strategies are followed, each with its own set of situational characteristics. A copy-cat scenario may be adopted where those same situational characteristics are followed for placement of branches in submarkets not yet penetrated by a competitive chain. Or an avoidance scenario may be followed where a new set of situational characteristics are adopted and followed in the expansion strategy, thereby avoiding direct locational competition with the competing chain and at the same time targeting niche markets overlooked by the competing chain.

Successful retail chains have followed any of myriad situational characteristics. Some of those characteristics are listed in box 7.4.

The analyst expects that if two retail branches in different locations have the same situational characteristics, then the two retail branches should have the same performance. This was discussed in chapter 4 in the context of Applebaum's analog method.

Box 7.4.   Example situational characteristics

- Locate branch in a tourist area
- Locate branch near a traffic generator such as a regional mall
- Position the branch on the homeward path and homeward-bound side of street, between the traffic generator and residential area containing highest density (see chapter 4) of primary target demographic population
- Locate branch as an outparcel at the highest order and highest end regional mall in the trade area
- Locate branch where the median household income averaged over a 40-mile trade area exceeds the median for the United States, and where there are at least 1.2 million people
- Locate branch next to a multiscreen theatre complex
- Locate branch in power center (see box 7.1), where there is shared parking for 250 vehicles within 800 feet of main entrance way
- Locate branch within one-quarter mile of freeway exit and with signage visible from the freeway, and where the daily traffic volume count exceeds 42,000 cars
- Locate branch within one-quarter mile of brand name competition
- Locate branch in historic downtown undergoing redevelopment as a fashion and entertainment center
- Locate only one branch in the highest order center of a rural county whose total population does not exceed 40,000 persons
- Locate branch where within a 10-minute drive (see table 2.2) there are over 90,000 households, projected population growth rate exceeds 4 percent, one-third of the population are between the ages of 18 and 24, household income is projected to increase by 10 percent
- Locate branch on an interior-of-the-block position along a street face, avoiding the difficult-to-access corner lot
- Locate branch as a street-side, high visibility outparcel, avoiding the loss of visibility from being a member of a strip center (see Box 7.1)
- Locate branch adjacent to other retail outlets with a compatible value platform and synergistically offering a good tenant mix
- Locate branch where there is available dedicated parking
- Locate branch where the average age of nearby commercial buildings does not exceed five years.

Numerous reasons may explain differences observed in performance, including management skill, pricing, product, costs, competition, and location.

Comparative branch performance measurements for a multibranch retail chain may be visualized with GIS. A thematic map might be drawn with a circle over the location of each branch. The circles might be color coded to represent branches that are above average in profits and below average in profits. The radius of the circle might be set proportional to the difference in the individual branch's profits and the average for the chain. Such visualization can bring to the attention of the analyst problems that may exist by region: a regional manager or regional taste preferences might be at fault. Such visualization will also identify submarkets that might be oversaturated or opportunistically available for greater penetration.

In many industries, multiple-outlet markets return less revenue per outlet than single-outlet markets. Higher-performing outlets may not have other outlets sharing their market. The analyst may therefore need to separate analysis into markets where single branches are located versus markets where multiple-outlets are located.

One of the key considerations listed in box 7.4 is traffic volume count. Daily and annual traffic volume counts for the past year and for the previous five years are available from a variety of data vendors listed in Thrall (2001). Transportation engineers at the state or county level collect the data. Private data vendors then assemble the locally collected data, geocode the position of the traffic measurement, translate the data into GIS ready format, and package the data on CD-ROM or the Internet. The analyst may then visualize the nearby traffic volume counts as a thematic map or perform further GIS-based analysis such as calculating the total traffic volume count within a five-mile radius of some coordinate.

Box 7.5 shows a set of factors that were reverse-engineered based on sites of Red Lobster restaurants and the most frequently occurring situational characteristics among the many restaurant branches. It is not known if real estate market analysts at Red Lobster Restaurants use any of the characteristics listed. Most multibranch retail chains hold their location strategies in close confidentiality with the concern that it may fall into the hands of a competitor, even though by the application of their strategy, their predominant geodemographic characteristics and other qualitative situational attractions can be calculated. Among the most important nondemographic situational attractions is traffic volume count.

Traffic volume may increase because of the presence of a traffic generator, such as a destination regional mall or multiscreen movie theater complex. Flow-through traffic along an interstate highway may increase local traffic volume and may also be used by an analyst as prospective customers of the retail outlet.

Once a situation has been determined, the analyst may evaluate, identify, and recommend specific sites. The difference between site and situation is largely a matter of geographic scale. Situational characteristics can describe phenomena over many hundreds of square miles. Site characteristics are descriptions of the immediate environment of the retail site, including adjacent tenants, visibility of the store from the street, and so on.

Some consumers tend to be biased toward multipurpose shopping trips. The effect is different for each industry and can even change with the cost of gasoline and the level of traffic congestion measured by flow-through of traffic volume count. Multipurpose shopping trips often begin by a household traveling to a traffic generator. A

Box 7.5.   Reverse-engineered situational characteristics for Red Lobster restaurants

---

- Average age 33–35 years of age
- 2.95–3.15 persons per family—probably two adults and one child
- No bias toward a particular housing value type
- Younger adults with a higher income level
- African-American households that have a income higher than the average African-American household for the MSA as a whole, but otherwise Red Lobster does not target specific African-American household income segments
- 50 percent to 79 percent of total population is white
- 40 percent to 49 percent of total population is African American
- 25 percent of renters are white, while 75 percent of renters may be minorities; young marrieds of any race may be the target
- Median housing value of $40,000–$69,999
- $20,000–$24,999 median household income
- $25,000–$29,999 median family income
- White per capita income of $12,000–$13,999

These characteristics were derived without input or communication with Darden Restaurants (Red Lobster) in any manner; these characteristics have in no way been confirmed or denied by Darden Restaurants (Red Lobster).

---

traffic generator is a primary destination drawing from a large trade area. Traffic generators include regional malls, large employment centers, large entertainment centers, tourist destinations such as a popular beach, and so on. Consumers may first travel to a regional mall, then to a restaurant, and afterward go home. To take advantage of multipurpose shopping behavior, a retail store may be situated between a traffic generator and the area containing the highest density of prospective customers. This is not a one-way benefit. Among the consumer's attraction to a traffic generator can be the ancillary businesses located along the travel path. An analyst for a multiscreen movie theater chain may evaluate the geographic pattern and density of supplementary businesses such as restaurants and retail stores that customers with its demographic profile are known to enjoy (see Thrall et al. 1996).

Step 5: Assess Market Penetration

In step 5 the carrying capacity of a business within a market area is calculated: the balance between profit margin, demand thresholds, and cannibalization between sister units of the same chain. In this way, the real estate market analyst can determine where to focus attention on real estate development and where to focus attention on asset disposition. This is invaluable for long-range strategic planning in corporate real estate.

Market penetration is the percentage of total market captured by the firm (see Ingene, 1984). A very low percentage of market captured may reveal an opportunity for the firm to expand in that market. A high percentage may suggest that there is little opportunity for further expansion in that market and that investment of additional resources may be better allocated to other markets.

But how is total market calculated? *Total market* can be the total population of the submarket, defined for example as the primary trade area (see chapter 4) of the retail branch. The geographic technologies used to derive these numerical values were presented in chapters 2 and 4. However, total market might be calculated based on the demographic characteristics of the targeted population. Most firms focus on a subset of the total population in their primary trade area. For example, based on the numbers presented in table 2.2, while the total population within a 10-minute driving time from some coordinate may be 91,787, all of that population may not be potential customers. The target population may instead be the 5593 households whose incomes range from $50,000 to $75,000 per year. The analyst must be careful not to double count if more than one demographic category is used. Most private data vendors will calculate and sell complex demographic counts on request, where many criteria must be met in order for the person or household to be included.

An alternative to population counts are revenue projections as a measure of the market. For example, CACI Marketing Systems (Arlington, Virginia) sells data for states, counties, Metropolitan Statistical Areas, demographic market areas, census places, ZIP codes, census tracts, and census block groups.[9] Among the data CACI provide are the number of people and the likely demand for a product or service by category of product or service. A firm could use its own revenue information combined with data provided by an appropriate data vendor to arrive at alternative measures of total market and market penetration (see Lee, 1989; Lee and Koutsopoulos, 1976).

Say that total market is measured as the total population of the Metropolitan Statistical Area (MSA). Market penetration, or market saturation, is then the ratio of annual guest count (total number of customers per given year) to the population of the MSA where the branch is located. The ratio can be as low as 0.0, meaning no population is being served. The ratio can also be well over 1.0, the numerical equivalent of serving the entire population more than once in a year. If the units are located in multiunit markets, then the guest counts for each unit would be aggregated. The smaller the resulting ratio, the less saturated the market with the particular firm. The market might be able to support one or more additional units. A high ratio identifies markets that might be saturated, meaning that no additional openings should be scheduled. A high ratio might also mean store closings would be appropriate. If competition is a relevant consideration, then the real estate market analyst will need to estimate guest count figures for competing firms and use those measurements in the calculations for market saturation.

However it is measured, market penetration is the proportion of the total market (total demand) captured by a particular retail chain. Market penetration is a measure of market presence and performance and competitive position. This information is necessary in answering the questions: Can a given geographic area support another unit of our chain? Are there too many units in this geographic area? Are there too few units in this geographic area?

Red Lobster Restaurants in Single Unit Markets

Ratio of 1990 Guest Count

to 1990 MSA Population

Ratio of 1990 Guest Count
to 1990 MSA Population

● 1.5 and above
◉ 1   to 1.49
◍ 0.5 to 0.99
○ 0   to 0.49

Note: Only those units open
before 1989 were included
in this analysis

Figure 7.7 Market saturation for single branch markets, Red Lobster Restaurant data for 1990 (from Thrall et al. 1998d).

The map in figure 7.7 shows the ratio of the 1990 guest count for Red Lobster restaurants by MSA to the MSA population. Only units in single-unit markets are reported in figure 7.7. Some Red Lobster restaurants in 1990 experienced guest counts that were equal to or greater than the population of the metropolitan area. Several restaurants were calculated as having ratios greater than 3.0; that is, the Red Lobster Restaurant served an equivalent number of meals to each person in the metropolitan area eating there at least three times. The regions that Red Lobster restaurants perform particularly well in are readily identified in figure 7.7. Those Red Lobster Restaurant units that were performing poorly are immediately identified as well; those along the Gulf Coast of Texas were performing well below the national average.

Step 6: Identify Geographic Markets for Expansion

To complete step 6, identifying geographic markets for expansion, the real estate market analyst must produce a variety of analyses that will refine potentially profitable markets and identify markets that are high risk. But not all basic analysis is prepared in the corporate headquarters GIS war room. The real estate market analyst must, with reports in hand, go into the field and visit the markets targeted for potential expansion or contraction. This and the foregoing steps are crucial to corporate real estate risk management. Both are discussed below.

Considering Both Unit Performance and Market Penetration Before going into the field, the universe of prospective submarkets where the retail firm may locate a branch needs to be reduced to a manageable number. Those sites that are most likely candidates for expansion should be identified to minimize the cost of site visits, increase productivity, and minimize error in possibly locating in a market that may not meet expectations.

The submarkets are identified that meet the target criteria presented in the above five steps. Reports prepared from the previous five steps reveal gaps in the market area coverage. The retail real estate market analyst will track the geographic pattern of competitors' branches and the degree to which submarkets have been saturated both by their client/employers and the competition. Saturated submarkets offer limited, high risk, or no opportunities for further expansion. Those saturated submarkets are subtracted from the universe of prospective submarkets; the residual submarkets are revealed to be candidates for market penetration.

The previous five steps must be considered holistically. The steps are to be used interdependently with one another and viewed together when making a judgmental decision. For example, the retail real estate market analyst should consider the results of step 3 in this chapter, assessing the relative performance of retail units, and the results of step 5, assessing market penetration. Putting steps 3 and 5 together, the real estate market analyst can determine in which regions/markets individual units are performing best or worst, and if the local population has accepted the product or service; which markets are saturated and therefore ineligible for expansion; and which markets are to be targeted for expansion. Steps 3 and 5 are assembled in table 7.6 and mapped using GIS in figure 7.8.

Table 7.6. Judgmental scenarios from combining steps 3 and 5, assessing market penetration and store performance

| | Step 5 criteria | |
Step 3 criteria	Large circle: market penetration is high	Small circle: market penetration is low
Large tack: unit performance is higher than average	(a) Excellent management and/ or location, but may not be sustainable, not a likely target for further expansion	(c) Target geographic market for expansion
Small tack: unit performance is lower than average	(b) Steady state; possible reduction in number of units might be apropriate	(d) Possible target for market expansion after further analysis including reevaluation of target situation demographics and site location strategy within the primary trade area and evaluation of unit management

Large and small tacks and circles are displayed on the map of figure 7.8. From Thrall et al. (1998e).

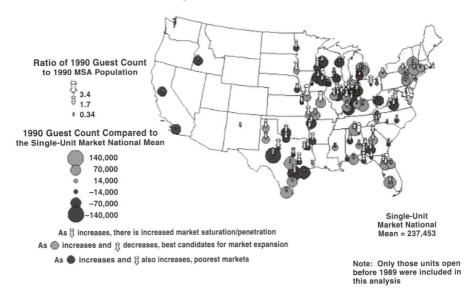

Figure **7.8** Multitheme criteria for identifying markets for expansion, steady state, or contraction, Red Lobster Restaurant data for 1990 (from Thrall et al. 1998e).

Four possible scenarios result when steps 3 and 5 are combined. Table 7.6 introduces the four management scenarios. The reference to the tacks and circles in table 7.6 links that table to the legend and interpretation of the map in figure 7.8.

Table 7.6 shows that, when a unit is performing higher than average for the group within the region (step 3) and the measure of market penetration in the geographic area is high, then there is a good likelihood that the manager of that particular unit is doing a superior job or that the population within the market area of the unit was particularly well suited for the product or service offered. This is shown in Table 7.6 as case (a). When a unit is shown to be performing lower than average for the group within the region (step 3) and the measure of market penetration in the geographic area is high, then that geographic market should not be considered for further expansion. Reduction in number of units might be appropriate. This is shown in Table 7.6 as case (b). When a unit is shown to be performing higher than average for the group within the region (step 3) and the measure of market penetration in the geographic area is low, then that geographic market should be one considered for unit expansion. This is shown in Table 7.6 as case (c).

When a unit is shown to be performing lower than average for the group within the region (step 3), and the measure of market penetration in the geographic area is low, then either (1) the local population within the submarket has not gained acceptance of the product or service offered, (2) the target demographics within the submarket was inappropriate; (3) competition has not been appropriately considered; or (4) the manager of the unit is doing an unsatisfactory job. Further expansion within the market area, and reconsideration of the submarket, depends on determining the cause for low performance. Further analysis is advised before the geographic market is considered for unit expansion This is shown in table 7.6, as case (d).

Single-unit markets are usually small towns with populations larger than required to sustain one unit, but insufficient to sustain more than one unit. Red Lobster restaurants located in multiple-unit markets have sales performances that are below those in single-unit markets. Therefore, single-unit and multiple-unit markets should be evaluated separately.

In figure 7.8, the multiple themes presented in table 7.6 are displayed on one map. In figure 7.8, 1990 guest counts for single-unit markets are compared to the national single-unit market mean and contrasted to the ratio of 1990 guest counts to the population of the MSA where the respective unit is located. The larger the size of the tack in figure 7.8, the greater the lever of market penetration or market saturation. The larger the circle, the better the individual unit performs compared to the to the national mean of unit performance. Figure 7.8 can be interpreted using table 7.6. As the circle size increases and tack size decreases, the existing unit is performing very well and the market is not saturated. This is the scenario of case (c) in Table 7.6. This GIS-based strategy then identifies the best candidate geographic markets for expansion. Market areas identified here as case (c) types include: Hartford, Connecticut, Albany, New York; and Lexington, Kentucky. After this strategy has brought these markets to the attention of the retail real estate market analyst, then he or she would evaluate the market using the criteria listed in box 7.3.

Going into the Field   This book has demonstrated the important and increasing role of desk-bound analysis, particularly desktop computer geography technology and related geographic analysis. However, this does not reduce the importance of the real estate market analyst traveling to the markets where one or more new outlets may be opened, closed, or relocated. The analyst must still drive around the city, visiting the prospective sites and making a visual assessment. The computer has not replaced traditional on-site evaluations (Thrall 1998a).

It is logically and technically possible that someday, perhaps even soon, field visits will be replaced with technology. There could be a "Star Trek" like "holodeck" where the analyst without leaving corporate headquarters may gain the same experience of actually driving and walking the prospective market and thereby arrive at the necessary qualitative assessment. The ideal holodeck would include a virtual reality of the city (and attributes of the city) projected to some future date. However, until that time, a personal onsite inspection remains necessary. Today, persons who have local knowledge and expertise should be invited to join with the analyst in his or her onsite inspection.

Before submitting a market analysis report to decision makers, the retail real estate market analyst must visit the prospective market. The qualitative evaluation of the market gained from that experience is combined with the technical reports. This information is then generally passed on to the corporate real estate department, strategic management, and the real estate deal maker. The real estate deal maker is responsible for combining the recommendations of the real estate market analyst with his or her knowledge and ability to negotiate a contract on a site (known as a *pad* in the industry). The real estate market analysis and the real estate deal making are generally distinctly separate operations. The reason they are kept separate is that a good real estate deal should not drive the location decision; instead, the deal should be struck within the constraints defined by the real estate market analyst. Also, the expertise, although overlapping, is manifestly different. Separating the two departments within the corporate organization thereby provides checks and balances.

Step 7: Judgment

Judgment is part of every step. And just as steps 3 and 5 were combined in step 6, each step requires appropriate and qualitative judgment. All the steps should be considered holistically; synergistically they must interact to improve judgment. However, each step is listed separately to allow for greater understanding of the analysis required of the step.

The below Associated Press story captures the essence of and underscores the importance of judgment. We can do wonderful things with the technology. We can routinely organize information along the lines of the six preceding steps. We can calculate gravity models. We can perform computational analysis like those presented in the earlier chapters and in the above six steps. We can do all this with unprecedented productivity and accuracy. However, if we do not use appropriate judgment, we are going to wind up headlong in the river.

> POTSDAM, Germany (AP) A German couple out for a Christmas drive ended up in a river, apparently because their luxury car's computer forgot to mention they had to wait for a ferry. The driver and his passenger were not injured in the accident, police said Saturday. Several companies sell computer navigators, some which are attached to dashboards and serve as electronic road maps. Some offer traffic updates and Internet connections. The German couple was driving Friday night when they came to a ferry crossing at the Havel River in Caputh. That information was never stored in the satellite-steered navigation system they were using. The driver kept going straight in the dark, expecting a bridge, and ended up in the water. (Thrall, 1998b)

The value that business places on real estate market analysis is that it improves decision making. Does the decision-making ability improve because of the analysis or not? Is the improvement in decision-making ability worth the expense? Although this must be decided on a case-by-case basis, the overwhelming trend is that estate market analysis is becoming increasingly sophisticated in its methods and more important as a tool for decision makers.

Four Micro Steps

As in the previous chapters, retail real estate market analysis also follows the four general conventional steps: identify trade area, estimate competitive supply, estimate demand, and produce report. Because multibranch retail chains today proceed with their corporate expansion decisions following a top-down procedure, the earlier and larger part of this chapter has focused on procedures relevant to that corporate real estate decision. It is at the micro level, after a market has been targeted for expansion, that the four-step procedure outlined in the previous chapters becomes particularly relevant.

Because the methods at the micro level have become so well developed for the retail side of real estate, they serve as an example for the other real estate product categories. Therefore, the procedures for identifying a trade area for a single retail unit were presented in chapter 4, as were the procedures for calculating competitive supply and demand. Some of the procedures introduced above in this chapter can be used at the micro level as well.

Concluding Remarks

More than any other real estate product category, retail real estate market analysis is more closely aligned with recent advances in geographic technology and adoption of geographic analytic methods. It was for this reason that the examples of chapter 4 were based on retail real estate. There are good reasons for this adoption and subsequent dependency.

The demand arises from large, multibranch retail firms that are rapidly expanding, making many development decisions each day. Each development decision is commonly a multimillion dollar investment. The company expenditures are largely concentrated in those branches, and the company's revenues originate from those branches. The location pattern of all the branches combined must make sense; the location pattern of the branches combine to add or detract value from the company.

The supply arises from the rapid innovation of geographic technology beginning for business applications primarily in 1990, with the release by the U.S. Census of the digitized street network known as TIGER/Line files, with improvement in desktop computer hardware, with standardization of operating systems interface, and with improved technology for distributing data such as the CD-ROM and the Internet. The new technology allowed for easier, more accurate, and routine assessments. The new technology allowed for the adoption of analysis such as that which has been presented here, but before this time was not cost effective to perform.

The background to retail business geography market analysis has been presented in this chapter. Cultures from around the world, for millennia, have been concerned with locating human activity at sites and situations that meet their objective criteria. Beginning in the western cultures in the nineteenth century, retail location analysis has become a complex routine performed by most multibranch retail chains. The procedures as discussed in this chapter are geographically arrayed top-down, beginning at the macro geographic scale of the globe, and proceeding downward to the evaluation at the very precise micro site-specific scale.

The macro scale analysis was divided into seven steps: identify area for store assessment, perform geographical inventory, assess the relative performance retail units, identify situation targets, assess market penetration, identify geographic markets for expansion, and make judgment. At the micro scale, the primary market area is calculated, followed by estimation of demand and estimation of competitive supply, and then compilation into a report. The methods to calculate trade area, demand, and competitive supply were presented in chapter 4 because, among all the real estate product categories, retail real estate market analysis is most highly developed, most complex and technically advanced, while at the same time the methods advanced for the retail side of the discipline have been adopted and used by most every other real estate product category.

8

Hotel and Motel

The key concepts, proceeding top-down, for market analysis for the hospitality industry are market segmentation, demand, and supply. Location or trade area comes into the analysis as an umbrella over these three concepts. Market niche and segmentation, demand, and supply are primary determinants to establishing the criteria for locating hospitality facilities.

Background

Whenever there have been sufficient numbers of travelers in search of food and shelter, some form of hostelry industry has arisen.[1] The Code of Hammurabi (1800 B.C.E) referred to innkeeping (Winfree 1996). In the western countries, as the Romans established an extensive roadway system, taverns and inns followed at strategically spaced locations. The Roman roads were used for military travel, trade and commerce, and pilgrimage and tourism. These are the primary reasons we use roads today. The early inns were largely run by religious orders. However, in Europe, as commerce grew in the fifteenth century, lodging as a commercial activity began to replace innkeeping as a charitable activity.

In the American colonial period during the seventeenth and eighteenth centuries, inns and taverns were an important part of commerce and cultural exchange. These facilities were designed after the inns and taverns of England, which were closely integrated into their communities. Inns and taverns did not intrude or disrupt the neighborhood; instead, they were thought of as being an integral part of the culture and activities of the neighborhood. Architecturally, early inns and taverns conformed to the look and feel of the surrounding neighborhood environment. Survivors of these early inns are the contemporary bed-and-breakfasts (B&Bs).

The term *hotel* arose early in the nineteenth century and was used to distinguish a greater level of commercial activity than an inn. Hotels offered food, drink, retail shopping, and lodging. Hotels were also more intrusive in their neighborhoods. Instead of less than 10 rooms that typified many inns of the era, early hotels contained as many as 200 rooms, and rose to 6 floors in height. Many nineteenth-century hotels were the tallest buildings in town. Thus, the hospitality industry began its first cautious attempts at market segmentation and diversification. Inns remained, but hotels offered an alternative experience via amenity differentiation.

In the latter half of the nineteenth century, the grand hotel was born, first appearing in the great cities of Europe (Rome, Paris, London), and afterward in New York City, followed by other large cities of North America. Soon thereafter, in every city and town where the threshold demand was met, there was a grand hotel waiting to offer the traveling public an ostentatious experience.

The hospitality industry was an early adopter of using more than location to attract their clients. Amenities were known to attract clients. In the hospitality industry this is known as *supply-induced demand*. William Flagler was an early proponent of supply-induced demand. Flagler used his fortune earned in the petroleum industry to finance an extensive network of railroads in nineteenth-century Florida. Although he could build and pay for the network of iron rails, there was little existing reason for anyone to use those lines in that undeveloped state. Flagler built a chain of ostentatious hotels strategically placed along his rail line. Florida winter weather, in combination with the amenity attraction of his hotels, induced demand for his railroad. Flagler even connected the Florida Keys with rail trestles that terminated at a Flagler hotel in Key West. Some Flagler hotels still operate as edifices to the past, such as in Key West and in Palm Beach. Other nineteenth century ostentatious hotels have been converted to public uses, such as housing colleges (in Tampa and Saint Augustine, Florida), and government offices (in Gainesville, Florida). Supply-induced demand arising from built and natural amenities remains as a significant niche of the twenty-first-century hospitality industry.

Before the streetcar, traveling even a few miles in the nineteenth-century industrial city was a time-consuming chore. Therefore, within every urban center there were strategically placed hotels and inns, some fitting unobtrusively into their commercial or residential district, and others ostentatiously towering over their surroundings. As transportation technology allowed the traveler more locational opportunities, the hospitality industry responded by further diversification of product offerings. Many of the small commercial hotels built from the late nineteenth century through the 1920s were subsequently converted into apartment accommodations, and in some neighborhoods, the flop house was born. As more people traveled long distances via automobile, the roadside hospitality industry became established. As in the days of the Romans, roadside accommodations were strategically placed between major origins and major destinations. The motel industry was thereby born.

The nation's motel industry had its origins along two major transportation routes in the era preceding the interstate. The mother highway of the nation, U.S. Route 66 (Wallis 1981) connected Chicago and Los Angeles. Later, U.S. Route, 301 and 441 connected the major tourist origins of the urban northeast and the winter resort and retirement destination of Miami, Florida. Along these routes, the motel industry was born.

In 1956, the U.S. Congress authorized funding for 42,500 miles of limited access highway, and the U.S. Interstate Highway System was born, which further changed the hospitality industry. The interstate changed travel patterns within the major cities and between U.S. cities. Before the interstate, urban hotels were largely concentrated near commercial nodes, particularly the central business district. The interstate led to suburban industrial and office parks. As office parks arose, the hospitality industry followed with further market segmentation aimed at business travelers and their suburban destinations. The synergy was recognized between office buildings, retail and entertainment centers, and hotels. A new form of multiuse was born by combining these land uses (more on this in chapter 9).

But not all travel was by way of the interstate. Post–World War II long-distance business travel shifted to the air. High-speed jet travel and dramatically reduced air fares allowed for more frequent business travel. Business personnel traveled to attend meetings and training sessions. The airport became a logical meeting place because all business travelers had to travel through that point. In geographic terms, the airport transportation center became what geographers refer to as an intervening opportunity. The airport became a destination capturing an increasing share of the business hospitality and business meeting room trade. The airport hotel was born.

Not to be outdone by the suburban airport, public–private partnerships arose in declining central business districts of major U.S. cities. Redevelopment efforts in many of cities centered on the convention center. The convention center serves to assemble many thousands of persons; these people stay in hotel rooms, eat meals, shop, and so on.

Diversification and market segmentation has in no way been left behind as a twentieth-century artifact. Greater and greater diversification and market segmentation remain as a paramount force of the hospitality industry. Business travelers increasingly travel with their families and look to multipurpose luxury business/recreational destinations. Weary business travelers, staying on their own, might choose a B&B for its restful home service and feel. Price-conscious long distance travelers along the interstates might instead look to budget accommodations with few amenities beyond access to the highway.

The above historical narrative of the hospitality industry reveals several dominant themes of the industry:

- Hospitality products will be increasingly diversified with greater market segmentation and focus on smaller and smaller niche markets.
- Some amenity-oriented niche markets respond well to supply-induced demand. In other words, build it and they will come.
- Location matters. But one location does not accommodate all niche markets. The location of an interstate off-ramp might be paramount to the success of a hospitality product aimed toward the price-conscious long-distance traveler, but it may have little importance to the destination-oriented, amenity-driven niche market. Target niche determines locational value.

The hospitality industry now faces two interrelated problems: a decline in occupancy rates and aging properties. Declining occupancy rates are a result of overbuilding in some submarkets. An overall occupancy rate of 85 percent was achieved in 1948. Post–World War II demand was high, and hospitality facilities could not readily increase in part due to the nation's priority of concentrating expansion of the housing supply. Since that time, the number of rooms has increased greater than demand, resulting in a decline in the average annual occupancy rate. By 1972, the occupancy rate had declined to 59 percent. As the twentieth century ended, an occupancy rate in the 60 percent range became the expected norm for the industry. Difficulties of arranging financing following the savings and loan crisis limited the increase in supply of lodging properties toward the end of the twentieth century. The result was what might be a temporary increase in occupancy rates to nearly 70 percent. These national rates differ from region to region and submarket to submarket.

The occupancy rate problem is linked to oversupply. Oversupply is created by both too great a rate of new construction, and too low a rate of removing obsolescent properties from the market. Travel today along old U.S. Route 66 or U.S. Route 301, and you will see that many of those original motels remain. To some they are historic structures. From the vantage of supply and demand, many of those substandard properties can compete only on the basis of price. "The problem with the hotel industry is not over development, but under demolition" (Winfree 1996, p. 9). The result is lowering of the rental rates in the submarket, as well as lowering of the occupancy rate for the submarket. A suburban Holiday Inn is thought today to have an economic life of about 30 years. Lower-end budget hotels are thought to have an economic life of 15 or 20 years. It is less costly to construct new hotels than to fully renovate budget or even middle-tier hotels. If the location remains desirable, then replacement via demolition and redevelopment might be in order. If the location becomes undesirable (see Stapleton Airport example below), then it might be in order to adapt to some other type of land use or to demolish and replace the motel with properties in another industry.

Market Niche

The hospitality industry targets the niche of the market they choose to operate in by way of their *value platform*. The value platform includes the product that the guest receives, including the design and accommodations of the room, the services and public facilities available, the situation of the facility within the surrounding environment, and other attributes that in combination add up to the experience the guest receives from staying at the facility. The value platform is what is hoped and intended for the guest to experience and place a value upon. In contrast, *market niche* is how the customer ultimately responds to the value platform that has been provided. Table 8.1 lists one categorization of market niches in the hospitality industry and provides some descriptive characteristics that might be used to distinguish niche markets by this categorization.

The categories of table 8.1 are not mutually exclusive and collectively exhaustive of every niche market for hotels and motels in the entire hospitality industry.[2] Within each category of table 8.1, there can be further segmentation, including high-end budget such as a Hampton Inn, versus low-end budget roadside sleeping station such as a Motel 6. And if the past is an indication of what the future may bring, there will be new hospitality flags identifying new niches, meaning ever-greater market segmentation. Table 8.1 is therefore an example of how segmented and niche-specific the hospitality industry has become. The hospitality real estate market analyst and his or her client must begin the project by clearly defining which market niche is to be targeted and how the particular value platform is to accommodate that. Focus groups representing the target market niche should be consulted.

The hospitality real estate market analyst must be cautious to not load the analysis in favor of a project by defining a niche so uniquely that there is no competition. There may be few prospective guests willing to pay for the particular narrow value platform. It is neither the real estate market analyst, nor the owner, nor the developer who designates the behavior of the market niche. Instead, they hope that the value platform they have designed serves to satisfy the desired market niche target. The customer guest recognizes, or not, the existence of the value platform; the customer places a value on that and thereby establishes the market niche. The guest ultimately distinguishes between niches, establishes the importance of the niche, and places their own value on the value platform of the development.

A hotel or motel may serve several market niches as presented in table 8.1. Crossover niches serve to hedge opportunities, to increase market share, and possibly to increase room occupancy rates. However, by broadening the value platform to accommodate several market niches, there might be an unintended increase in competition from the other players in the new niche. Also, guests in one niche are not willing to pay for amenities demanded by guests of another niche that they do not place high value on. Examples of crossover niches in the hospitality industry include:

- Suburban office park-oriented hotels that serve the weekend getaway market and business traveler market that is traveling with family, thereby combining business requirements with family time
- The casino, combining the value platforms of gaming and family entertainment and hospitality accommodations
- The amusement park, combining the value platforms of family entertainment, hospitality accommodations, and business conferences.

The more diversified the value platform and the broader the market niche, the greater the prospective market. The larger market might submerge influences from any narrowly defined niche, making the project less susceptible to market downturns that would otherwise result from a too narrowly defined niche. However, hospitality demand is a constantly moving target. As markets, transportation, and technology change, then so, too, do niches change. The ability to accommodate crossover niches might lessen. Markets may change because of changing geodemographics and changing environments. A once thriving hospitality environment serving a large-scale traffic generator might decline because of changing conditions confronting the traffic generator that are outside the control of the hospitality industry.[3]

Changes in technology, especially the phenomenon of the Internet, continue to change the market. Virtual reality via the Internet might substitute for some business travel. What was once a locational advantage might drift into locational obsolescence, or the remote, amenity-rich destination might drift into having a locational advantage. But ever-changing diversification might bring new market niches as substitutes. Airports are broadening their value platform beyond being mere transportation terminals.

Table 8.1. Example market niches in the hospitality industry

Type	Location	City size	Facility scale	Quality	Amenities	Example
Convention hotel	Downtown or major edge city commercial node.	Top-tier metropolitan regions	400 or more rooms in one facility, with several thousand rooms available in adjoining facilities. Extensive meeting space and dining facilities available.	High-quality facilities generally less than 20 years old or renovated to be of same as new quality. Quality accoutrements include contemporary international design themes and near-new features including beds, wall decorations, and so on.	Usually adjacent to large, city-owned convention centers. Include exercise room, pool, lounge, banquet facilities, and meeting rooms in a range of sizes. Seating for banquets normally accommodates 1,000 persons.	Top-tier convention cities include Los Angeles, New York, Chicago, Dallas, Philadelphia, Atlanta, and Washington, DC.
Mixed niche, business convention and family holiday	World class: amusement parks, ski resorts, seaside beaches	1.25 million, and with airport hub; quality airline facilities attributable to the high traffic volume generated by the amusement parks	400 or more rooms in one facility, with more rooms available nearby. Meeting space and dining facilities available on site or nearby.	High-quality facilities generally less than 20 years old or renovated to be of same as new quality. Quality accoutrements include contemporary international design themes and near-new features including beds, wall decorations, and so on.	Adjacent to high-amenity natural or built environment location, such as world class water, amusement parks, and skiing. Seating for banquets normally accommodates 400 persons with a range of many restaurants nearby.	Hilton Hawaiian Village, Waikiki Beach, Hawaii; Disney World, Orlando, Florida; Vail and Beaver Creek, Colorado

Type	Location	Airport access	Size	Quality	Market/Draw	Examples
Executive conference center	Large or midsize city downtowns, airports, or secluded areas with resort-like accommodations	Cities serving as airline hubs	Large to midsize, 200–400 rooms	High-quality facilities generally less than 20 years old or renovated to be of same as new quality. Quality accoutrements include contemporary international design themes and near-new features including beds, wall decorations, and so on.	Adjacent to midsize city-owned convention centers with aggressive marketing program or providing convention center facilities on premises. Include exercise room, pool, lounge, banquet facilities, meeting rooms ranging from large to small.	Hilton Head, South Carolina; Vail and Beaver Creek, Colorado; Denver International Airport
Destination natural environment resort hotels	Lifestyle environmental amenity areas such as seaside or ski resorts	Normally must be within an hour drive from an airport served by one or more major airlines	Large to midsize, 200–400 rooms, including a nearby cluster of hospitality accommodations catering to a variety of niches	Range of high and medium quality. High-quality facilities ranging from less than 20 years old to historic opulent facilities renovated to be equal or superior to highest international standards. Moderate quality facilities, generally older, that have not remained competitive with newer development and were not of the opulent category when new.	The surrounding environment such as the mountains or beach might be the client pull, and/or the hotel architecture and amenities offered either add to the draw or be responsible for the draw. Golf and tennis can be a major draw if development is world class and is combined with world-class natural environment.	Marco Island Marriott Resort and Golf Club, Marco Island Florida (near Everglades on Gulf of Mexico); Palm Springs, California; Scottsdale, Arizona; Maui, Hawaii; Vail, Colorado; Napa and Sonoma, California; Hilton Head, South Carolina

(continued)

Table 8.1. (*continued*)

Type	Location	City size	Facility scale	Quality	Amenities	Example
Business destination economic base hotels	Single-purpose trip generated by significant local economic base	Midsize, usually not a top-ten city	Midsize, 160–225 rooms	Business class facilities, generally not more than 20 years old. Older properties upgraded to equal that of newer facilities. Onsite breakfast accommodations and small meeting rooms with quality restaurants nearby.	Adjacent to midsize convention center or shuttle service between hotels and convention center. Attractive historic restored downtown with restaurants nearby ranging from moderately priced good restaurants to superior quality restaurants	Huntsville Alabama, Marshall Space Flight; Washington, DC; Silicon Valley, California
Destination urban-built environment hotels	Historic downtowns, high urban amenity with high order shopping, nearby high cultural entertainment	Large, midsize to small cities. The urban built environment is the draw	Midsize	Range of high, medium and budget quality	Historic areas are nestled within pre-auto-era urban landscapes, providing for attractive walking environments. Medium to small size meeting rooms to serve small business meetings	Miami South Beach, the fish market in downtown Seattle, historic Santa Fe; Savannah, and Charleston, New Orleans, Broadway of New York City, shopping in Dallas, the historic ambiance of Boston
Roadside sleeping stations	Visible from and adjacent to high traffic volume interstate freeways.	Situated a day's drive between a high population origin and a high demand destination.	Ease of access is paramount. The only restriction on small city size is availability of labor to manage and service the hotel	Medium to lowest budget	Ample parking, a perception of a safe environment for guests and their vehicles, nearby chain restaurants and gas stations, perhaps a pool and hot tub	Low budget end: Motel 6, Motel 8, EconoLodge, High budget end: Quality Inn, Fairfield Inn, Comfort Inn, Ramada Inn, Holiday Inn

Type	Location	Size	Class	Description	Examples
Urban-based entertainment centers	High profile amusement parks and sports stadium districts.	Very large to midsize, often cross-marketed with convention center based hotels	High to medium business class	Situated within easy access to amusement parks or sports stadiums, often situated among spillovers from those traffic generators.	International Drive, Orlando, Florida, situated among spillovers from Disney World; hotels associated with the Superdome of New Orleans
Bed & breakfasts, inns, resort condominium rentals	Historic districts situated within built environment with high quality urban amenities, or access to special natural environment such as skiing or world-class water and in attractive rural areas	Usually small but might include historic renovated destination resort	Highest to medium	If situated within a city, the accommodation is normally in an attractive urban walking environment historic district. If near skiing or world class water environments, generally is walking proximity to the amenity, or publically provided shuttle service. If in a rural environment, generally is an historic structure with access to significant world class environmental amenities.	Sweetwater Branch Inn, Gainesville Florida; Redstone Castle, Redstone, Colorado; Fryemont Inn, Bryson City, North Carolina; owner-rented condominiums in Vail, Colorado.

Because many people are at the airport, and traffic volume count is a primary consideration for retail location, then airports might become important retail centers. Hence, a large, ostentatious, multiuse office-retail-hotel facility might require an airport location.

The hospitality industry is confronted with continually changing markets, and the hospitality real estate market analyst must anticipate those changes. A downtown may decline in its ability to attract conventions. A once booming resort area may become rundown, then revived as in the case of Miami Beach's South Beach. A new resort development in a previously undeveloped area might open to tourism, competing with other destinations nearby or, in a global economy, competing with destinations many thousands of miles away.

For the hospitality real estate market analyst, analysis begins with conceptualization of market segmentation—namely, identification of niche and value platform. After the niche or niches and the value platforms are identified and decided upon, the relevant trade area, competitive supply, and demand are calculated. These concepts are then integrated to present to the client a composite picture of room absorption rate, which in this context means occupancy rate, expected revenues, and number of guests.

Establishing a Trade Area

The overriding theme of the hospitality industry is increasing market segmentation, and that includes trade area delineation. A proposed destination resort hotel may draw its prospective guests from a national or even international market. A proposed roadside sleeping station (table 8.1) may draw its prospective guests from the traffic passing by the interstate exit up to one-half mile away. An historic inn like Redstone Castle, in Redstone, Colorado, might draw its guests from touring, middle-aged Harley-Davidson motorcycle riders; the upscale female riders and passengers are the ones making the decision as to where to stay, and which route to take (Schmitt 1999). The demographic profile of the guests of that inn will match the demographic profile of Harley-Davidson motorcycle riders. And the riding habits of those riders serve to determine the trade area for the Inn. The primary trade area would be defined as within a 2-day motorcycle ride, or 500 road miles, and the secondary trade area would extend an additional 250 road miles away, based on the typical distance ridden by a touring motorcycle rider. The demographic profile of Harley-Davidson motorcycle riders is given in table 8.2 and illustrates:

- The high level of niche targeting and market segmentation practiced by the hospitality industry
- That the trade area depends on the targeted niche, market segmentation, and the value platform being provided
- That the target demographic characteristics can be approximated in a first estimation by known demographic characteristics of other goods that guests are known to consume (Applebaum estimated this by noting the make, model, and year of vehicle that patrons were observed driving and parking in the retailers' parking lots).

Estimating a trade area for a unit in the hospitality industry, then, is not a simple application of a single geographic formula. Instead, a variety of niche-specific rules

Table 8.2. Redstone Castle Inn's demographic profile based on guests' association with a product (Harley-Davidson motorcycles) whose demographic characteristics are known

	1987	1988	1989	1990	1991	1992	1993	1994	1995	1996	1997	1998	1999
Gender													
Male	98%	96%	96%	96%	95%	95%	93%	93%	91%	91%	93%	93%	91%
Female	2%	4%	4%	4%	5%	5%	7%	7%	9%	9%	7%	7%	9%
Median age (years)	34.7	34.6	34.6	36.7	38.5	38.4	41.6	42.1	42.5	43.6	44.6	44.4	44.6
Median household income ($K)	38.4	40.0	44.7	47.3	50.5	53.7	61.9	65.2	66.4	68.5	74.1	73.6	73.8

From http://www.harley-davidson.com/company/investor/inv_18.asp.

of thumb must be applied. Table 8.3 lists the same niches within the hospitality industry as 8.1 and provides a descriptive narrative of how the trade area of each niche might be approximated. In some niches, the guests may be drawn from international, national, or regional markets. However, this might be irrelevant in determining the true trade area.

The true primary trade area in many instances is best thought of in context of the geographic distribution and proximity of the phenomena that draws guests to the locale. For example, although a ski resort hotel within the destination natural environment resort hotel niche of table 8.1 might draw its guests from an international market, an individual lodging's effective trade area is the nearby ski slope. Vail, Colorado, attracts roughly 1.5 million skiers each year. About 40 percent of those skiers are day trippers from Denver, 100 miles away, and other front range origins. So, about 800,000 skiers on Vail's slopes are prospects as hotel guests. Most skiers place a premium on proximity to the slopes and après ski amenities. Thus, the effective trade area is the distance skiers are willing to travel from the slopes. Vail provides a free ski shuttle service that extends only within the town of Vail. The ski shuttle service then defines the competitive trade area, from which competitive supply and demand are drawn.

Estimating Demand

Say the target market niche is a roadside sleeping station (see table 8.1 and 8.3). People stay at a roadside sleeping station because it is convenient to their transportation route. There are adjacent retailers offering goods and services required by the traveler, including restaurants and gasoline stations. The rooms are clean, quiet, and the room rates are low. To broaden the niche, the motel might provide a pool that the guest (and especially their children) might use after the day's long drive. An estimation of demand for a roadside sleeping station begins with the identification of the important situational elements and evaluation of whether the important situational elements are present at the location. Table 8.4 presents a hypothetical evaluation matrix of situational characteristics for a low-end budget roadside sleeping station as defined in table 8.1, whose trade area is defined in table 8.3.

Table 8.3. Trade areas of illustrative niches within the hospitality industry

Type	Trade area
Convention hotel	A convention hotel in North America would be viewing the combined U.S. Canadian, and Mexican markets. The target population from which prospective conventioneers would be drawn would include the number of persons who are members of any national association that meets annually. Estimates of the total numbers of persons can be obtained by compiling a list of target associations, or from private data vendors.
Mixed niche, business convention and family holiday	This niche would also view the combined U.S. Canadian, and Mexican markets. The target population from which prospective conventioneers would be drawn would include the number of persons who attended conventions in competitive locations. Estimates of the target population can be obtained by compiling a list of target associations; their attendance and location of meetings are generally listed on their Internet sites. Or, the data can be purchased from private data vendors.
Executive conference center	This niche may be the combined U.S. Canadian, and Mexican markets, but more likely limited to the national market. The target population from which prospective conventioneers would be drawn would include the number of persons who attended conventions in competitive locations, which can be obtained from local chambers of commerce, or the data can be purchased from private data vendors.
Destination natural environment resort hotels	The primary market (chapter 4) for a U.S. hotel would be national, with an important secondary market that would be international in the North American Hemisphere. A tertiary market would be from developed countries elsewhere in the world, drawing from that population whose incomes and lifestyle segmentation profiles (LSPs) match that of North American consumers. By mapping the density of persons who match the LSP of the destination hotel, important submarkets can be identified, target populations estimated, and populations targeted.
Business destination economic base hotels	The prospective guests would be drawn from a market that is national, with a possibly important secondary international market. However, as the purpose for the guests' visit depends on the demand generated by the local economic base, then the true clients are the nearby firms and institutions that provide the purpose for the visit. Therefore, it is more appropriate to use a rule-of-thumb drive-time measurement, such as fifteen minutes. The demand then can be estimated following Applebaum's analog procedure where the independent variables might include the number of firms by standard industrial classification code, the number of employees by firm, whether the firm is a headquarters or branch plant, and so on.
Destination urban-built environment hotels	The prospective guests would be drawn from a market that is national, with the possibility for larger more important destinations of an important secondary international market. The purpose for the guests' visit depends on the demand generated by the local urban-built environment. The true clients then might be envisioned as those players in the local urban-built environment, such as restaurants, theaters, acreage of historic district, and so on. It would be appropriate to use a rule-of-thumb drive-time measurement to measure trade area, such as fifteen minutes. The demand then can be estimated following Applebaum's analog procedure where the independent variables might include the number of restaurant chains by category of restaurant (four star, three star, etc.), by square footage of shopping, and so on, within the trade area.

Table 8.3. (*continued*)

Type	Trade area
Roadside sleeping stations	The prospective guests would be drawn from a very large market, but one that is regional. The true primary trade area may extend from an interstate exit, only a quarter mile, and a secondary trade area an additional half-mile. Therefore, it is appropriate to use a rule-of-thumb drive-time measurement to measure trade area, such as five minutes. The demand then can be estimated following Applebaum's analog procedure where the independent variables might include the nonlocally generated traffic volume count and other variables.
Urban-based entertainment centers	The prospective guests would be drawn from a market that is national, with the possibility for larger more important destinations of an important secondary international market. The purpose for the guests' visit depends on the demand generated by the local urban-built environment. The true clients then might be envisioned as attendees of sports arenas, and so on. It would be appropriate to use a rule-of-thumb drive-time measurement to measure trade area, such as fifteen minutes. The demand then can be estimated following Applebaum's analog procedure where the independent variables might include: the average annual patronage of the activity; if public conveyance is available to shuttle prospective guests from the site to the activity; and frequency and timeliness of the conveyance.
Bed & breakfasts, inns, resort condominium rentals	The prospective guests would be drawn from a market that is national, with the possibility for larger more important destinations of an important secondary international market. The purpose for the guests' visit depends on the demand generated by the local urban-built environment. It would be appropriate to use a rule-of-thumb walking-time measurement to establish trade area, such as fifteen minutes brisk walking, or 1.5 miles. The demand then can be estimated following Applebaum's analog procedure where the independent variables might include the acreage of historic district, number and square footage of retail stores, number of chairs by restaurant category, and frequency and timeliness of public transportation.

Niche definitions of this table correspond to those of table 8.1.

Table 8.4 is an hypothetical description of the important nearby situational characteristics for a roadside sleeping station, or low-end budget facility. Table 8.4 is structured in a format that lends itself to model development in the tradition of the Applebaum analog procedure (chapter 4). Situational characteristics 1–9 include complementary items that hospitality guests may desire including interstate access and the existence of ancillary goods and services. Traffic volume count is a measure of potential market from which guests are drawn. Situational characteristics 10–through 13 include traffic generators that may or may not be important to this category of facility; however, their existence might be calculated in the analog procedure as increasing the occupancy rate by a few percent. Examples of potential traffic generators include a regional mall, a university, an airport, and a regional hospital.

The variables in table 8.4, SC_i for n hypothetical situational characteristics, where $i = 1 \ldots ,n$ are defined as dummy terms, are equal to 1.0 if the characteristic is present and 0.0 otherwise. Also included is the driving distance separating the proposed site and the situational characteristic.

Table 8.4. Hypothetical situational characteristics (SC) for a low-end budget roadside sleeping station

Characteristic	Present		Driving distance (miles) between characteristic and proposed site
	Yes	No	
SC1: At least two competitive gas stations	SC1 = 1	SC1 = 0	0.75
SC2: At least two competitive fast-food restaurants (McDonalds, Wendy's, Burger King, etc.)	SC2 = 1	SC2 = 0	0.5
SC3: At least one breakfast restaurant (Waffle House, Denny's, etc.)	SC3 = 1	SC3 = 0	0.25
SC4: More than four budget sit-down chain restaurants (Cracker Barrel, Sonny's BBQ, Bob's Bigboy, etc.)	SC4 = 1	SC4 = 0	0.85
SC5: At least one middle-end sit-down chain restaurant (Red Lobster, Olive Garden, TGI Fridays, etc.)	SC5 = 1	SC5 = 0	1.25
SC6: At least one grocery store (Safe-way, Vons, Kash & Karry, etc.)	SC6 = 1	SC6 = 0	1.75
SC7: Interstate access nearby	SC7 = 1	SC7 = 0	0.25
SC8: Interstate freeway traffic volume count between 20,000 and 30,000 cars each day	SC8 = 1	SC8 = 0	0.25
SC9: Interstate freeway traffic volume count exceeds 30,000 cars each day	SC9 = 1	SC9 = 0	0.25[a]
SC10: Regional mall or power center present with > 600,000 square feet	SC10 = 1	SC10 = 0	1.35
SC11: Regional or national university present with > 20,000 students	SC11 = 1	SC11 = 0	4.25
SC12: Regional or international airport present serving > 4,000 passengers per day.	SC12 = 1	SC12 = 0	8.0
SC13: Regional hospital present with > 100 beds	SC13 = 1	SC13 = 0	3.25

[a] SC8 and SC9 may both be 0.0, but both cannot be 1.0.

Measurements of traffic volume count should be considered in the context of the targeted niche market. Roadside sleeping station target demographics require those persons counted in the traffic volume count to be included only if they are traveling long distances (i.e., not commuting locally to work or shopping). Thus, the real estate market analyst must separate out those persons driving long distances from local commuters. Various demographic data vendors resell traffic volume counts collected by state and local departments of transportation. Using spatial interaction models (see chapter 4), the traffic volume count attributable to local in-place demographics can be estimated. The difference between gross observed traffic volume counts and those estimated from in-place demographics is the count of population from which guests will be drawn. Some private data vendors will sell traffic volume counts of non-local traffic.

Deriving Target Demographics

Table 8.2 displays the demographic profiles of persons who use a particular product and are known to frequent an inn. By analogy, the demographic or LSP profile of the inn might on first estimation be interpreted as being the same as for the unrelated product (Harley-Davidson motorcycles) the guests are known to consume.

Given an address, a geographic information systems (GIS)-based LSP product can calculate latitude and longitude coordinate, as well as an LSP index (see table 4.6), for that address. The addresses can be obtained from guests who have stayed in branches of a chain or purchased from competing chains. Using the GIS-based kernel method (see chapter 4), the density of prospective guests can be calculated and maps drawn of that density. The real estate market analyst should observe a significant density of prospective guests within a single day's or several days' drive from a prospective site.[4]

A Hedonic Demand Model

In chapter 4, regression analysis and the development and calibration of hedonic models were discussed in the context of Applebaum's analog procedure. Based on that discussion, and the above, a hedonic model for the hypothetical roadside sleeping station can be specified.

The dependent variable can be yearly or seasonal revenues, R. Alternatively, the dependent variable might be specified as number of guests, NG, or occupancy rate, OR. The hospitality real estate market analyst might estimate hedonic models based on all three dependent variables.[5] The independent variables on first estimation using stepwise regression might include all those variables listed in table 8.4. Therefore, a hedonic model for yearly or seasonal revenues might be specified as:[6]

$$R = A + \beta_1 SC1 + \beta_2 SC2 + \beta_3 SC3 + \ldots + \beta_{13} SC13. \qquad (8.1)$$

More complex hedonic models can also be measured. The advantage of more complex models is that they might yield a greater level of precision in estimation. For example, it might be known that the importance of a breakfast restaurant declines with distance from the motel. Geographers know this as the distance decay phenomena. Therefore, in addition to the presence of, say, a Waffle House Restaurant, the proximity of the Waffle House might be hypothesized to be important as well, with the revenues of the budget-end roadside sleeping station declining the farther the Waffle House is from the motel. In other words, the value of the regression coefficient, β_3 in the hedonic model of equation 8.1 might depend on D_3, the distance of the Waffle House from the motel. We can then hypothesize that the regression coefficient for the Waffle House variable is.[7]

$$\beta_3 = \beta_{0,3} + \beta_{1,3} D_3. \tag{8.2}$$

Substituting equation 8.2 into 8.1, we have:

$$R = A + \beta_1 SC1 + \beta_2 SC2 + (\beta_{0,3} + \beta_{1,3} D_3)SC3 + \ldots + \beta_{13} SC13, \tag{8.3}$$

or

$$R = A + \beta_1 SC1 + \beta_2 SC2 + \beta_{0,3} SC3 + \beta_{1,3} D_3 SC3 + \beta_{1,3} D_3 SC3 + \ldots + \beta_{13} SC13. \tag{8.4}$$

$\beta_{1,3} D_3 SC3$ is known as an interaction term and allows for the interdependence between both the presence of a Waffle House and its proximity to the motel.

Estimating Supply

The straightforward component of supply is establishing the number of competitive facilities already built within the market area. The business geographer engaged in hospitality market analysis has a variety of public sources for competitive supply.

Sources include the yellow pages of a telephone book and various hotel/motel guides such as from the American Automobile Association (AAA). Data are discussed more fully in Thrall (2001). Geographic coordinates of competitive facility locations can be calculated based on facility address. After the potential competitive supply has been geocoded, the final determination of competitive supply is estimated by commanding the GIS software to count the number of competitive facilities within the defined trade area (see table 8.3). If number of rooms per facility is known, then the GIS software could be commanded to calculate the total number of rooms in the specified trade area.

Room counts can be generated by visual inspection, published hotel guides, telephone calls, or can be purchased through a private data vendor (Thrall, 2001). Room counts can be estimated from facility records, since motel chains by brand have fairly standard count of rooms. Occupancy rates can be estimated by telephone calls, visual inspection, or purchase though private data vendor. However, the hospitality market analyst should be skeptical of the accuracy of occupancy data and should verify occupancy rate with visual inspection and personal contacts.

Pipeline data for the hospitality industry are available for a few of the larger U.S. markets (Thrall 2001). Pipeline data are required for the estimation of future supply of competitive product. However, hotels are more ubiquitous than the largest 100 U.S. markets normally covered by pipeline data vendors. An interstate exit at Perry, Georgia, is today unlikely to be included in published pipeline data. But Perry might be a good location for a roadside sleeping station motel. The hospitality real estate market analyst must visit the prospective market, speak with the local chambers of commerce, financial institutions, and builders, review zoning files and building permit applications, and follow up with conversations with hotel developers and managers. Competitive nearby interstate exits must be evaluated in a similar manner. This muddy boots determination of competitive supply is an arduous task, but it is important because the construction of one unanticipated large facility can possibly oversupply a trade area with rooms.

Competitive supply is not necessarily a bad thing for the individual firm. The benefits of agglomeration might overwhelm the disbenefits from increased competition. Consider again the case of the hypothetical roadside sleeping station. The existence of a few competitive motels at the same interstate exit might generate the agglomerative presence necessary to attract guests. The guests among all the available beds, including the competition, provide the threshold demand to support services that are desired by one's own guests, but which might not be provided based on one's own guest counts. Therefore, as the number of beds increases (NB) in the trade area, then demand for one's own beds increases. However, beyond some level, a further increase in NB will diminish occupancy rates. This describes a polynomial function, which, following the above procedures, might be modeled as:[8]

$$SC14 = \beta_{1,14}(NB) + \beta_{2,14}(NB)^2 + \ldots + \beta_{n,14}(NB)^n \qquad \text{for } n > 2. \qquad (8.5)$$

The right-hand part of equation (8.5) is then added to the hedonic equation of (8.4).

Any number of reasonable expressions can be added into the hospitality hedonic model and thereby capture the various site, situation, and competitive factors that combine to create number of guests, occupancy, and revenues.

Compile Report and Present Analysis to Client

A hospitality real estate market analysis describes the overall project within the wide context of the economic, geographic, and social forces that are shaping the hospitality industry. The final report should include a presentation of the real estate analyst findings on market segmentation (niche and value platform), trade area, supply of competitive product, and demand estimates for the product. The report should place the project within the social, economic, geographic, and even political context of the community. The report should also include a statement of the goals and objectives of the client as the analyst understands those goals. The analyst then combines his or her demand and supply analysis to estimate hotel room absorption, including best and worst case scenarios of occupancy rate projections.

Concluding Remarks

The market analysis for hotels and motels share some of the methods presented for the retail real estate product. The hospitality industry is closely associated with many of the components that create demand for retail. The analyst must identify the value platform and hoped-for target niche for the hospitality product. That in turn sets limits and defines the relevant trade area. As with retail, multibranch hospitality chains can construct hedonic models that can be used to estimate revenues and occupancy rates. Also as with retail, the multibranch hospitality chain should devise a geographic expansion strategy along the lines of that discussed in chapter 7.

9

Mixed Use

With perhaps the exception of building a new town, mixed-use (MXD) development requires the most complex real estate market analysis.[1] As with the structural organization of the preceding chapters on real estate products in this book, this chapter will begin with a background of the real estate product type. A background of MXDs is necessary to understand those developments that are already in place across the North American landscape.[2] Some MXDs have been successful and others have been dismal failures. A goal of this chapter is to describe and explain the instruments hypothesized to make an MXD successful. Some MXDs are approaching their functional age of obsolescence—25 or 50 years old. They may require new real estate market analysis to guide their redevelopment and that redevelopment must be executed in the context of how they originated. The background coverage, contemporary notions of trade areas, demand, supply, and report, for MXDs are presented.

Background

What Is a Mixed-Use Development?

To be defined as an MXD, the real estate project must have three components (Schwanke 1987):

- Three or more significant revenue-producing uses (such as retail, office, residential, hotel, and/or entertainment/cultural/recreation), which in well-planned projects are mutually supporting

- Significant physical and functional integration of project components (and thus a relatively close-knit and intensive use of land), including uninterrupted pedestrian connections
- Development in conformance with a coherent plan, which frequently stipulates the type and scale of uses, permitted densities, and related items.

Each of the above concepts is discussed below.

Three or More Significant Revenue-Producing Uses

Many real estate projects have multiple uses. However, MXDs as defined and discussed here must have at least three major revenue-producing uses. These uses should be nontrivial. In other words, if retail space is one of the mixed uses, then that retail space should have a trade area beyond the mere project site. In most contemporary mixed-use projects, retail, office, residential, and/or hotel facilities are the primary revenue-producing uses. Other revenue-producing uses of MXDs might include sports arenas and convention centers, performing arts facilities, and museums. The importance of the multiple uses at a single project is their interdependence: each should create demand from a trade area, and that demand should spill over to add to the demand of the other uses at the site. The aggregate demand from such interdependencies should be greater than each use would generate on its own.

The definition of MXD, requiring three or more significant uses together in one development, generally restricts true MXDs to projects of significant scale and of large regional impact. However, smaller-scale MXDs have become more common in recent years, but they still usually exceed 100,000 square feet. Smaller projects are generally highly integrated into the local built environment, such as a downtown historic district (Thrall et al. 1996). Larger projects run into several million square feet. A rule of thumb on size is that all MXDs must achieve a minimum critical mass to become imbedded into the public consciousness and for market penetration to occur from each category of use. The goal of proper MXD design is to use size and diversity of uses to create a significant new place on the urban landscape. MXDs therefore generate, capture, and benefit from, the creation of a sense of place.[3]

Physical and Functional Integration

An MXD must achieve a significant physical and functional integration of project components. Normally this requires, and in turn benefits from, an intensive use of land. Pedestrian byways should interconnect all components of the project. The successful MXD must be designed to be a pedestrian-friendly environment, and means are provided to conveniently move pedestrians from one end of the MXD to the other, as well as provide connections to the nearby environment external to the MXD. This integration and connectivity can take many physical forms. Larger MXDs will comply with all of the below features:

- Vertical integration, where there is an ease of upward and downward movement, including both visual and physical integration. The sense of place must be designed to connect physically and visually upper and lower spaces. MXDs might be combined into a single building, with larger MXDs being megastructures and comprising many buildings.

- Spatial organization of key project components should serve to integrate the various components, so that all benefit from public spaces.
- Interconnection of project components should be accomplished through pedestrian by-ways, including sidewalks, escalators, interior walkways, enclosed corridors, underground concourses, aerial bridges, and public spaces of gallerias, retail plazas, and mall areas.

There must be a pedestrian focus to the overall design of the MXD to achieve the desired synergies between project components and to create a sense of place. The pedestrian focus must include the geography of pedestrian movement, both within the MXD and between the MXD and the surrounding built environment. Without the pedestrian focus, the MXD will not be successful.

The second criterion of physical and functional integration ultimately is what distinguishes a successful MXD from other real estate projects, even though those other real estate projects have three or more significant revenue-producing uses. If the key components are not integrated in the manner described above, then the project cannot be called a true MXD. For instance, lower-intensity, land-extensive developments such as suburban master-planned communities and business parks may meet the criterion of three or more revenue-producing components, but the geography of those components result in less frequent interaction between uses and a greater dependence on automobiles to connect the key components of the project. Such developments are not true MXDs.

Large suburban business parks may include light industrial uses, office buildings, retail and restaurant space, hotels, health clubs, golf courses, daycare centers, and so on. Their geography is that of a low-density, extensive land use, with key components connected primarily by way of private automobile. A geographic dispersion of the key components and lack of public pedestrian conveyances is not conducive to good pedestrian circulation. Some master-planned communities incorporate multiple uses in their design, such as shopping centers, office buildings, hotels, and residential and recreational uses. These projects are generally large scale and may be several thousand acres in size. However, their low density begets a dependence on the private automobile for internal connectivity, as well as for connectivity to other nearby built environments. Therein lies the distinguishing characteristic that separates MXDs from most other real estate projects. Non-MXD developments generally create spaces whose key components are separate and missing the full potential benefit that interrelatedness can bring.

MXDs generally follow a coherent development strategy and plan that is established at the outset of the project. MXD master planning demands a greater diversity of specialized participation from developers, business geographer market analysts, architects and land planners, property managers, financial analysts and capital/financing sources than is the case of single-use projects.

The development plan for an MXD is generally composed of a collection of materials such as market studies, a development program, land use and building configuration plans and models, working drawings, cost estimates, feasibility analyses, and financing plans. Large MXDs may require many years of planning, development, and absorption. During that time, the development program may need to change in response to changes in the local or national economy and unanticipated positive or negative market response to various components of the project. These options may be interpreted as best-case and worst-case scenarios, and each scenario should contain a distinct development program. Each distinct development program should be anticipated as part of the initial planning of the MXD. Each separate development program requires a separate real estate market analysis.

History of MXDs

MXDs draw on numerous experiences with urban development. Medieval towns have served as models and downtowns of the era that preceded the automobile have served as models. For example, the oldest continuous inhabited European settlement in the United States is the city of Saint Augustine, in northeast Florida. The footprint of the preindustrial historic city was used as the model for the Haile Plantation MXDs city center in Gainesville, Florida, in the late 1990s. Haile Plantation is a 2,000-acre multi-use, suburban development. It contains various sorts of housing, recreation (golf, tennis, swimming), and at the center is a true MXD. The Haile Plantation MXD consists of high-density retail, offices, and housing. The MXD is integrated to the surrounding multiuse development by way of pedestrian/bicycle paths and an automobile roadway network. Lot widths in the MXD are narrow. Retail or office space occupies the first floor, with housing above. Parking and garages are relegated to back alleys. The density of the Haile Plantation MXD, width of streets, and curvature of streets, are a copy of those in historic sixteenth-century Saint Augustine.

At the other end of the scale according to size, complexity, and regional impact scales is the Rockefeller Center MXD in New York City. Rockefeller Center is an example of a large, very dense, big-city MXD. Its creation was largely driven by demand for office space. However, the developers integrated numerous office buildings with retail, recreational, cultural and entertainment (i.e., ice rink and Radio City Music Hall) facilities. Rockefeller Center has been very successful and has become a prototype for large MXDs. It was built in roughly two stages (1931–1940) and (1946–1975). Rockefeller Center has continued to be developed, and even redeveloped, in ensuing years. At the start of the twenty-first century, the rentable area of Rockefeller Center is 17 million square feet, and there are 21 skyscrapers on its 24 acres. All the buildings in Rockefeller Center have been designed to create a sense of place and to have a positive spatial relationship with one another. A quarter of the land area of Rockefeller Center has been left in public open space.

Common characteristics contributing to the success of the very large Rockefeller Center and the comparatively small Haile Plantation MXD are

- Pedestrian accommodation, and vehicle routing separate from pedestrian byways
- Retail space serving populations within the development, and drawing populations from a larger trade area
- A one-management solution for organizing development in the MXD space, allowing an integrated approach to design and land use, so that new structures are harmonious and contribute synergistically to the overall design concept.

An MXD Missing the Target Criteria

Not all MXDs have been successful (Schwanke 1987). Century City is a large multiuse development about 20 driving minutes west of downtown Los Angeles. The development of Century City began in 1961 and was completed about 1990. The developer was a wholly-owned subsidiary of Alcoa Aluminum Company. The original plans were for Century City to have 12.5 million square feet of office space, 1300 hotel rooms, 3,300 residential units, and 880,000 square feet of retail space, theaters, a hospital, and parking.

On the one hand, the project was remarkable for its conceptualization, which began in the 1950s. However, the project was also a product of its era and location, and therein lie the reasons for its failure. Although the project accommodated the automobile, its accommodation created significant physical barriers and loss of functional integration of the components of the project. Specifically, the project's design required

- Wide boulevards separating large super-blocks speeding the flow-through of vehicular traffic
- Pedestrian traffic connecting the super-blocks, uninterrupted by small streets and alleys
- Auto parking restricted to be remote from desired places of access
- Wide green spaces more than 80 percent of the land separating the key components of the project.

Although each of these items may be laudable in different contexts, within the context of the MXD they created a loss of integration of the key components of the projects.

MXDs Hitting the Target Criteria

The office market drove the creation of both Century City and Rockefeller Center, although they were very different in design (Schwanke 1987). The importance of the office markets driving the development of MXDs cannot be overestimated, both in large and small-town CBDs and suburban areas. Office land use can most easily exist independent of other land uses within the development. However, the demand created by office developments for other ancillary uses is a development opportunity and a significant factor in the market analysis of an MXD (see Thrall et al. 1995, Thrall et al. 1996). Retail land uses and residential land uses have less frequently been the cornerstone of success of an MXD; still, retail land uses and residential land uses have played a significant role in supporting MXDs, particularly in suburban areas. The well-designed MXD, regardless of the land use that has been the dominant reason for its development (office, retail, housing), is capable of internalizing and benefiting from the market development forces that arise from its development.

Establishing a Trade Area

From the vantage of the real estate market analyst, and their obligations to the client, the MXD is the most complex form of land use. Applebaum's analog procedure (see chapter 4) is the cornerstone of most real estate market analysis. The analog procedure has been used in every chapter on market analysis of a specific real estate product. A key ingredient for using the analog procedure is that performance is repeatable if the geography of site and situation are repeated, and that requires a homogeneous, stylized real estate product. For example, if identical restaurant design, food, menu, and layout

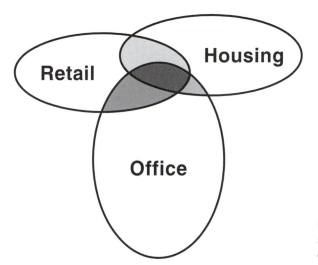

Figure 9.1 Real estate market analyses required for a three-component MXD.

is located in the identical geographic context but at another location, then the performance of the restaurants should be repeated.

The added complexity of performing real estate market analysis for an MXD comes from the very thing that sets them apart and makes them desirable: their heterogeneity and the synergistic interaction that arises from their key components. Instead of one restaurant, they have many. Instead of a single traffic generator such as a regional mall, they may also have traffic generators such as offices and hotels, as well as housing for permanent residents. The mixed-use chapter comes after the other presentations of real estate market analysis by specific product both because of its complexity and because the market analysis is a composite for each type of land use that the MXD contains.

Figure 9.1 is a Venn diagram depicting the array and potential interaction of land use types for an MXD. The real estate market analyst must calculate the trade area for each of the key land uses within the MXD, as well as the effect (either positive or negative) from their pairwise or multiway interaction. In this hypothetical example MXD, the three key land uses are office, retail, and housing.

Separately, office, housing, and retail trade areas are calculated as presented in the preceding chapters for the respective land use type. Then the interaction effects must be considered. Unless the net interactive effects result in larger trade areas or greater demand within a trade area than would be the case for each land use type separately, then the complexity and greater risk of an MXD development is not warranted. The interactive effects (shaded areas of figure 9.1) would include:

- Retail/office: A greater retail demand from office employees and office visitors leads to a greater demand for office space because of the amenities offered by retail such as restaurants.
- Housing/office: Office workers require housing, and some might place a high premium on distance minimization between home and office (see chapter 2).
- Retail/housing: The greater the density and proximity of housing, the greater will be

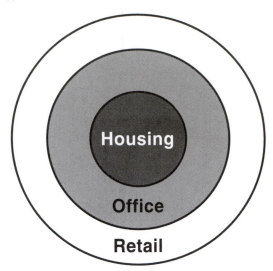

Figure 9.2 Hypothetical trade areas
for pure land use types within an
MXD.

the demand for retail items sold. For some, excellent access to retail is a predominant
consideration for their choice of housing.

- Retail/office/housing: a multiplier effect (see chapter 4) may ensue, thereby providing
  for an even richer and more interesting built environment than would be provided by
  any pair of land uses or any land use on its own.

The trade area must be calculated separately for each of office, retail, and housing, as
illustrated in figure 9.2. As with the discussion centered on figure 9.1, either the
demand within a trade area or the trade area itself will be greater; this is necessary
to justify the added complexity and greater risk of the MXD. Figure 9.2 depicts trade
areas for each land use type independent of the interactive effects that may occur
within the MXD. Considering the interactive effects, because users of the MXD are
being drawn from the larger retail trade area shown in figure 9.2, then housing and
office are expected to piggyback onto the larger retail trade area.

Estimating Demand

The demand must be estimated for each of the separate key land uses of figure 9.1.
Subsequent to that, the change in demand resulting from the interactive effects between
the land use types must be calculated. If the aggregate demand including the interactive
effects is not greater than the demand for each of the key land uses evaluated separately, then the complexity of the MXD development might not be justified. Among
the interactive effects that might increase demand are:

- Employment: The presence of white collar and professional work force in the MXD
  offices creates a favorable climate for both housing and retail.
- Income: Because the MXD offers a pedestrian-friendly environment, and as office
  workers choose to reside at the MXD, then the MXD resident household might opt not

to own a car, or to purchase one car instead of two or more. That savings might translate into a greater proportion of MXD household income spent on more luxurious housing and more upscale retail.

- Target demographics: Not all households will find an MXD suitable to their lifestyle. Those who choose to reside at an MXD will have identifiable lifestyle characteristics. Those characteristics may be derived from comparable MXDs elsewhere, through surveys. The business geographer then determines the location, number, and density of persons with that same lifestyle profile in the target market area.

Estimating Supply

As mentioned in the discussions of figure 9.1, the business geographer must estimate the current supply of competitive product and competitive product in the pipeline, both for each land use as separate components and jointly with the various offerings. For some markets in the United States, the current supply and the supply in the pipeline are provided for both MXDs and the components of MXDs by a variety of data vendors. See Thrall (2001) for further discussion of data. If commercial data are not available, then the analyst will need to perform onsite data gathering like that in the example for hotels in chapter 8.

Compile Report and Present Analysis to Client

The MXD real estate market analyst's description of the overall project within the wide context of the economic, geographic, and social forces that shape the city is more important in this real estate product category than in any other. The reason for the MXD to exist comes from that interdependence, and the greater independencies that the project will create. The final report should include a presentation of the business geographer's findings on interdependencies and the resulting overall value platform, trade area, supply of competitive product, and demand estimates for the product. Not only should the report place the project within the social, economic, geographic contexts, it should also place the project within the political context of the community. The report should also include a statement of the goals and objectives of the client, as the analyst understands those goals. The report should reference the difficult managerial task of organizing the various components and the possibility of failure of the project if one of the key components does not produce or does not perform. The analyst then combines his or her demand-and-supply analysis to estimate absorption rate of the various components of the project. Because of the intense interdependencies, best-case and worst-case scenarios of the MXD project must be included.

Concluding Remarks

The business geographer must quantitatively and qualitatively evaluate the market for an MXD. Included among the quantitative findings are the trade area, the demand, and competitive supply. Included among the qualitative findings are the more difficult-to-calculate components that arise from the interdependent nature of the MXD development—namely, the synergistic effects between each of the key land uses of the MXD, as well as market trends such as those discussed in chapter 3. The business geographer must consider the greater risk arising from the increased complexity of the MXD development and compare and contrast that to the returns that would arise from separate-use developments.

10

Getting Started

The business geographer performing market analysis for real estate should become skilled in the advances of geographic technology, as well as geographic and real estate analysis and procedures. And the client should become skilled in judging the analyst's work. In this context, the eighteenth-century poetic essay by Alexander Pope (1688–1744) is appropriate (see box 10.1). The left column is particularly relevant to the analyst practitioner, while the right column is particularly relevant to the client who is making his or her judgmental decision.

The client, whether an investor, financier, or developer, should know enough about business geography and real estate market analysis to correctly understand the evaluation and report, know which questions to ask of the analyst, and know how to translate the report into correct judgment. The client making the judgmental decision should not have his or her vision clouded by details of the choice made for which data source to use for population projections, nor should the judgmental decision be steered off course by the choice between which desktop GIS software to use. Instead, the client has other considerations, such as How do I select and work with a business geographer performing market analysis for real estate projects? and When and how should a business geographer consultant be used? This chapter gets the reader started in these tasks.

Consultants

Financiers, investors, developers hire business geographers to provide a variety of services, including choosing the appropriate data, software, and methods to use, and

Box 10.1. Excerpt from *Essay On Criticism*

A little learning is a dang'rous thing;
Drink deep, or taste not the Pierian spring:
There shallow draughts intoxicate the brain,
And drinking largely sobers us again.
Fir'd at first sight with what the Muse imparts,
In fearless youth we tempt the heights of arts,
While from the bounded level of our mind,
Short views we take, nor see the lengths behind,
But more advanc'd, behold with strange surprise
New, distant scenes of endless science rise!
So pleas'd at first, the tow'ring Alps we try,
Mount o'er the vales, and seem to tread the sky;
Th' eternal snows appear already past
And the first clouds and mountains seem the last;
But those attain'd, we tremble to survey
The growing labours of the lengthen'd way,
Th' increasing prospect tires our wand'ring eyes,
Hills peep o'er hills, and Alps on Alps arise!

A perfect judge will read each work of wit
With the same spirit that its author writ,
Survey the whole, nor seek slight faults to find,
Where nature moves, and rapture warms the mind;
Nor lose, for that malignant dull delight,
The gen'rous pleasure to be charm'd with wit.
But in such lays as neither ebb, nor flow,
Correctly cold, and regularly low,
That shunning faults, one quiet tenour keep;
We cannot blame indeed—but we may sleep.
In wit, as nature, what affects our hearts
Is not th' exactness of peculiar parts;
'Tis not a lip, or eye, we beauty call,
But the joint force and full result of all.
Thus when we view some well-proportion'd dome,
(The world's just wonder, and ev'n thine, O Rome!')
No single parts unequally surprise;
All comes united to th' admiring eyes;
No monstrous height, or breadth, or length appear;
The whole at once is bold, and regular.

Alexander Pope, 1711. *Essay on Criticism*. London.

rely on their professional skills of execution and ability to complete and present the report in a manner that will improve their judgment. The business geographer brings objectivity, professionalism, and both broad and specialized experience with similar projects.

How should a business geographer be chosen? First, the prospective client should decide what project(s) the analyst is to evaluate. The type of projects a business geographer might be engaged to work on include;

- Determining the highest and best-use for a given site
- Selecting the location and evaluating the viability at that location for a specific type of development

- Constructing an expansion strategy and location strategy for individual outlets of a new or existing chain of retail stores.

The client should decide whether outside expertise should be sought or whether the real estate market analyst functions should instead be performed in house.

Contracting out for business geographer services is an approach used by Home Depot. Home Depot has decided that their business is retailing of items for home maintenance and home improvement. Business geography is peripheral to that business activity. Developing an in-house team of real estate market analysts is not within their business plan. The decision to contract out means that they need not deal with the questions of how an appropriate staff of analysts are to be assembled or how that staff is to be monitored and evaluated. In-house capabilities could become costly beyond the mere wages of the employees. Contracting out may benefit the firm by tapping experts that have a wide range of experiences that might benefit the firm. Consulting firms regularly tap their well-developed contact network for temporary expertise required to execute a project.

In contrast, Blockbuster Entertainment has developed in-house business geography expertise. A firm that has its business geography in-house requires a sufficient volume of activity to justify the daily activity and expense. An added benefit of having in-house business geography staff is that "location, location, location" can more easily become integrated into the larger managerial procedures of the company. In-house business geographers come to know and understand the corporate culture. They can interpret the changing needs of the corporation in the fast-paced retail world. They can be proactive in the type of reports required. Management can integrate their reports and expertise into the larger managerial operations of the firm. Instead of isolating the business geographer as a contracted-out specialty service, business geographers and the benefits their knowledge brings to the company become an integral part of the company. When managerial positions open in other divisions of the firm, the business geography staff might be tapped to fill the position because of their broad corporate knowledge and job experience in developing the locational value platform of the company. The business geography division then can serve as a farm to fill higher-level managerial positions. A downside for any firm that follows this approach is that, although knowledge of the firm might be extensive, close integration with the larger corporate culture may limit objectivity.

Whether a company develops in-house capabilities or contracts out for that expertise, the firm must have management expertise that can evaluate the product received from the business geographer before it is too late and a legacy of bad decisions has been established.

After the contracting out or in-house approach has been decided upon, consultants must be found or staff hired. How are these real estate market analysts to be found? Usually this is done by asking associates who are known and trusted for referrals. Ask other companies whom they have used, how they felt about the work, or how they found appropriate staffing. Look in directories of consultants and professional organizations. Network through trade and professional associations. Contact members of the board of directors of those professional organizations; current and former membership of a board of directors documents that an individual has respect of his or her peers and a broad contact network that can be tapped into. If new university graduates are sought, develop a relationship with key faculty of geography and real estate departments which have renown faculty in the field of business geography. Key faculty today keep in close touch with former students through e-mail. Those former students may be prospective candidates for newly opened positions. Appropriate university faculty can be determined in the same process as outlined above for new employees.

From the vantage of the prospective client, an added benefit of an Internet query is that many consultants have information on their web pages. From the vantage of a consultant, a well-informed client is preferred to one who is not knowledgeable. Less initiation time is normally required if the client is well informed. The consultant need not take a knowledgeable client down the learning curve to the point where the client can make a decision as to what they want the consultant to do, what characteristics they should be seeking in a consultant, and if they want to hire the consultant or not. Consultants are wary of prospective clients who may use them to merely gain knowledge of what their needs are and then perhaps perform the analysis in house by unskilled staff. Many consultants will charge an up-front fee that covers the costs of educating the client.

When the potential consultants are narrowed down to a handful, then select the best consultant for your firm. Interview each prospective consultant. Ask each consultant for a proposal. Their proposal should include a statement of services to be provided, a timetable to deliver those services, and the estimated costs.

Part of narrowing down prospective consultants (or employees) is an evaluation of their qualifications. Included in this evaluation is the documentation listed in box 10.2.

Once a consultant has been selected, then write a contract that addresses the services being provided, schedule, fees, and personnel that will be involved in the services. The contract should include a mechanism for adequate and periodic accounting of the work to be performed. The contract should include a mechanism for terminating the contract, which will protect both client and consultant. The client should ascertain the reasonableness of the fees proposed. The client and consultant should review all documents that reflect the proposed relationship.

The above information is relevant to the person choosing to become a business geographer and perform real estate market analysis. It sets forth what is required to be successful as a practitioner, whether as an independent consultant or as an employee for a large corporation.

Software and Data

As demonstrated in the preceding chapters, real estate market analysis is inseparable from business geography and geographic technology. Geographic technology is estimated now to be a $40 billion per year industry. $1.5 billion of that is from the sale of software and data. The remaining amount includes wages and earnings of persons engaged in consulting or as employees such as real estate market analysis or otherwise engaged in GIS applications for business or government.

The real estate market analyst must make choices as to which GIS software and data to use. A comparative evaluation of software is available in Clapp et al. (1997),

---

Box 10.2. Components for the evaluation of business geographers' and real estate market analysts' credentials

---

- Expertise and experience of working on projects similar to the proposed project
- Education credentials
  - Highest degree of educational attainment (generally, the higher the degree, the better the qualifications if the other categories on this list also reflect superior credentials)
  - Relevancy of subject area of university degrees to profession
  - How recently credentials were attained, and if subsequent training or activity in other categories on this list reflect currency of knowledge and up to date technologies and procedures
- Professional credentials and professional accolades
  - Member of a professional organization's board of directors
  - Member of an editorial board of a publication relevant to the profession
  - Publications within the publications and journals of the profession
  - Awards received by the profession
- Independent assessment of the person's qualifications by means of
  - Widely enjoyed reputation in the profession
  - Client and/or colleague references
  - The person's commitment to real estate market analysis, including to the particular type of real estate product that is being proposed

---

and list of data is available in Thrall (2001a,b,c). I recommend that the reader access information on data and software via periodicals on the topic, including *GeoSpatial Solutions* (*www.geospatial-online.com*), publications of the American Real Estate Society (*www.aresnet.org*), including their *Journal of Real Estate Literature* and *Journal of Real Estate Research*, and various publications of other professional organizations such as the Association of American Geographers (*www.aag.org*).

Just as there are consultants in real estate market analysis, there are also consultants in GIS who make recommendations on which software and data to use. Recommendations will differ depending on needs and scale of operations and whether the recommendations are made for an individual analyst or the largest global corporation requiring an Internet enterprise solution.

I had compiled a comprehensive annotated bibliography of data vendors relevant to business geographic analysis for real estate and intended to include it as an appendix to this chapter. However, the appendix easily added another 20% to the size of this book. And the data industry had already changed during the writing of this book. *The Journal of Real Estate Literature* reviewed the proposed appendix and chose to publish it in its June 2001 issue (Thrall 2001a). The categorization for the annotated data sources appears in Thrall (2001a) and box 10.3. The categorization has become the starting point for a regular data categorization by the American Real Estate Society;

Box 10.3.   Organization of data resources for business geographic real estate market analysis

1. Market-level economic base research sources
   1.1 Population
       1.1.1 Historical counts
       1.1.2 Current estimates and forecasts
       1.1.3 The economy
             1.1.3.1 Employment and other economic indicators
       1.1.4 Other economic indicators
             1.1.4.1 Major employers
             1.1.4.2 Labor productivity
             1.1.4.3 Unemployment rate
             1.1.4.4 Employee wages salaries and benefits
             1.1.4.5 Educational attainment
             1.1.4.6 Cost of housing
             1.1.4.7 Taxes and government expenditures
             1.1.4.8 Bond (Credit) rating
             1.1.4.9 Transportation
             1.1.4.10 Public policy issues
             1.1.4.11 Natural hazards

2. General real estate supply-and-demand sources
   2.1 Real estate forecasting reports
   2.2 Real estate market reports
   2.3 Specialized real estate sources

3. General real estate supply conditions
   3.1 Permits
   3.2 Starts
   3.3 Agricultural land

4. Data sources for a single property type
   4.1 Office
   4.2 Retail
       4.2.1 Retail supply and demand: Population, households, income, and competition
       4.2.2 Retail sales, rents, and expense benchmarks
       4.2.3 Retail supply
       4.2.4 General retail real estate reconnaissance
   4.3 Residential (nonapartment residences)
       4.1 Residential supply
       4.1 Residential sales volume and prices

(continued)

Box 10.3.   (*continued*)

4.4 Apartment rents
   4.4.1 Vacancy rates and homeownership rates
4.5 General residential reconnaissance
4.6 Industrial
   4.6.1 Industrial demand (Employment)
   4.6.2 Industrial supply
4.7 Hotels
   4.7.1 Hotel basic data sources
   4.7.2 Hotel reports

5. International—miscellaneous

From Thrall (2001a)

---

new additions of annotated data sources will appear either in *Journal of Real Estate Literature*, or on the American Real Estate Society web site. See *GeoSpatial Solutions* for an alphabetical listing of data sources, categorized by availability from government or commercial data vendors (Thrall, 2001b,c); the alphabetical listing is also available on *GeoSpatial Solutions* web site.

Conclusion

In this chapter, some of the nuts-and-bolts of the profession have been introduced with an aim of getting the work underway. Necessary for getting started is knowledge of the geographic technology industry, especially its software vendors and data sources. Some of the data suppliers are government agencies, and those data are generally available at no charge via the Internet. Private data vendors provide, for a fee, their data, which are distributed either over the Internet or by CD-ROM.

The improvement of the business decision is the raison d'etre for the profession. Practitioners must then be opportunistic and grasp new theory, new methods and practices, and new technology to serve and satisfy this need. The ultimate evaluation is whether judgment has been improved by an amount that exceeds the cost of the analysis. Perhaps we in our field should have a banner on our computer monitors that reads: "By judgments improved and decisions made, you shall know us."

Notes

---

1. For discussion of the importance of diversification of a real estate portfolio, geographic and otherwise, see Mueller and Ziering (1992).

2. Advice on pricing is generally delegated to certified MAIs (members of the Appraisal Institute). Increasingly, MAIs are adopting business geographic technologies and procedures to arrive at comparative pricing estimates. An example of such procedures is provided in chapter 5 on housing. The market analyst may assemble price information created by MAIs and not necessarily take on the responsibility or expertise for creating this information themselves. For a discussion of market analysis in the appraisal process, see Mueller and Wincott (1995); for problems of automating appraisal and market analysis, see Worzala et al. (1995).

3. The knowledge required to perform professional-level market analysis, financial feasibility, manageability, and investment risk is sufficiently great as to require specialist professionals in each of these areas.

4. The importance of general theory, like that presented in chapter 3, as well as general knowledge that comes from descriptive geography, like that presented in chapter 2, are invaluable in honing one's intuition about market forces that shape the geography of a city. This qualitative analysis from an experienced and knowledgeable specialist is an invaluable component of the final report to the risk manager and decision maker.

5. See Thrall (2001) for a discussion of data available to the business geographer/real estate market analyst.

6. There is ongoing discussion about increasing the frequency of updates by way of cooperation with the U.S. Postal Service and local governmental units.

7. See the discussion of Prisoners' Dilemma in chapter 2.

8. For discussion of the role that physical as well as psychological barriers can have in determining trade areas, see Morrill, et al. (1988).

9. For discussion of the importance of perception to real estate, see Golledge and Thrall (2000).

10. The business geographer/real estate market analyst, leaves the estimation of the cost of developing the site to a specialist in building construction cost estimation. Likewise, while certainly acknowledging that inflation can change profitability estimates, they leave such calculations of the impact of inflation (Pyhrr et al. 1990, Wurtzebach et al. 1991) to the financial analyst.

11. For discussion of real estate product types and their definitions, see Anikeeff and Mueller (1998).

Chapter 2

1. Lifestyle segmentation profiles can be used to specify the geographic location and extent of a housing submarket and a neighborhood.

2. A ratio of housing submarkets to office submarkets consistently defined and calculated across urban areas could reveal interesting characteristics of urban areas by urban size, as well as identify those urban markets likely to increase their number of office submarkets.

3. The model of urban land use proposed by Homer Hoyt in 1939 (see also Hoyt 1933, 1966), was that market forces would bring about distinct sectors of industrial, commercial, and residential land use, with low- and high-class residential areas radiating out from the old town center. Hoyt's model relied on the importance of radial transport routes and how those routes in turn determine the pattern of land use.

4. For discussion and an example of the role of GIS in Community Reinvestment Act analysis, see Thrall et al. (1995).

5. The consumption theory of land rent (Thrall 1987) is summarized in chapter 3 of this book. Transportation enters the general theory by way of cost that affects households' budget constraints. Also, transportation can be evaluated in the context of a "utility shifter," with utility diminishing with increases in the time required for commuting.

6. See Yeats (1990) for discussion of economic base and cycles.

7. This is similar to Hardin's (1968) "tragedy of the commons." One's welfare, and consequently one's decisions, are dependent on the actions of others.

Chapter 3

1. Von Thünen's well-known formulation for land rent has led him to be credited as the first of the "marginalists" school of thought in economics. Marginalists support the reasoning that profit maximization occurs where marginal revenue is equal to marginal cost. For further discussion of von Thünen, see Wheeler et al. (1998), Hall (1966), Dunn (1953), and Chrisholm (1965).

2. There have been many theories proposed to explain urban land values and land uses since the seminal contributions of Ricardo and von Thünen. Many of these theories are consistent with one another, and yield very much the same results, though they have been developed using different methodologies and different premises.

Chapter 4

1. This and related sections of this chapter are based on publications by Thrall and del Valle (1996a,b,c, 1997a,b,c).

2. Customer spotting techniques have subsequently been used by Ghosh and McLafferty (1987), Moloney (1989), and Rogers and Green (1978), among others.

3. Applebaum's analog method has been demonstrated by Drummey (1984), Ghosh and

McLafferty (1987), and Rogers and Green (1978). Drummey and Rogers and Green were among the first to combine Applebaum's analog method with multiple regression models; this concept is reviewed later in this chapter.

4. For discussion of the geographical differences of these locations, see Thrall et al. (1995, 1996).

5. There are various econometric techniques that can deal with this issue, including the Chow test, as well as the use of dummy variables, where a value of 1.0 indicates, say, a downtown location, and 0.0 otherwise. Advantages of the dummy variable approach are that the data can be pooled and retained in one data set, and the t values of the regression coefficients will reveal if the hypothesized geographic difference between the two environments is statistically significant.

6. Rogers and Green (1978) were the first to introduce the analog regression approach to retail site evaluation. Also see Drummey (1984); Douglas (1949); Epstein (1978); Ferber (1958); Goodchilan (2000).

7. This presentation of the gravity model is based on Thrall and del Valle (1997a). They followed the numerical example presented by Haynes and Fotheringham (1984).

8. The now-classic Huff model grew out of a collaborative and synergistic effort during one of geography's most important intellectual eras: the beginnings of the quantitative revolution. The Department of Geography at the University of Washington had in attendance some outstanding graduate students who went on to found the quantitative revolution: Brian J. L. Berry, Richard Morrill, Duane Marble, Arthur Getis, and Bill Bunge. These students had weekly brown bag lunches. Then-graduate student and business marketing major William Huff learned of these informal gatherings. He brought to the gathering a question how to calculate a geographic trade area. Out of the collaborative interaction came what is now referred to as the *Huff model*.

9. Calibration of the spatial interaction model (i.e., the assignment of values to the parameters of β and the other weights of model) is necessary for accurately estimating market areas, and particularly the calculation of cannibalization between retail centers. For methods of calibration, see Fotheringham and O'Kelly (1989) and Williams and Fotheringham (1984).

10. This section is based on Thrall (1999b, 2001). Also see for further discussion Moudon and Hubner (2000).

11. Dodge Pipeline data contain data for the past eight quarters. It is available by subscription via the Internet. Included as part of the "Real Estate Decision Tools" package is software that provides SQL-like (structured query language) ability to search, sort, tabulate, aggregate, export, and print information on projects that meet search criteria.

12. As with F. W. Dodge's data product, Comps.Com provides software that allows for SQL-like operations, as well as automatic report generation. Among the automated reports are market trends by real estate product category, such as apartments; the reports provide analysis at the Metropolitan Statistical Area and county geographic levels.

13. Reference to these data products here and in Thrall 2001, does not indicate support or preference by the author or Oxford University Press for these data products.

14. This section is based on Thrall (2000b).

15. Woods & Poole Economics Inc. specializes in long-term economic and demographic projections. Woods & Poole's database for every county in the United States contains twenty-year projections, and more than 550 variables.

16. County population growth is a function of both projected natural increase and migration that result from local economic conditions. Woods & Poole claim 10-year population projections to have accuracy within 13.9% for counties and 5.8% for states.

17. This section is from Thrall (1998b) and Thrall and McMullin (2000a). See also Batey and Brown, 1995.

18. CensusCD+Maps includes the complete U.S. Census STF3 (A, B, C & D) data files. The STF data files include demographic information at the block group, tract, place, MCD, Zip

code, nation, region, division, state, county, MSA, Indian reservation, and 104th congressional district levels. U.S. Census block-level data are available also from the same data vendor, under the product name CensusCD Blocks.

Chapter 5

1. For a more in-depth explanation of Hagerstrand's pioneering work and discussion of related literature, see Morrill et al. (1988). See Thrall et al. (1993) for a general theory and application of spatial diffusion to real estate market analysis, including forecasting housing absorption by small submarket. Also see Berry, 1972.

2. The penetration of retail franchises into a market also follows this same process, as discussed by Morrill et al. (1988). Also see Brown 1975, 1981.

3. In another related study, Thrall et al. (1995) used the same database to describe the timing and spatial pattern of development for housing, office, retail.

4. For a more advanced presentation of calculation of absorption rates based on the logistic S curve and spatial diffusion, see Thrall et al. (1993). The article is available on the Internet at the web site for the *Journal of Real Estate Research, www.aresnet.org*. The URL for the article is *http://business.fullerton.edu/journal/pdf/vol08n03/v08p401.pdf*. Also see Alves and Morrill (1975); Boyce (1966); Brown (1975); Brown (1981); Casetti (1969); Colenutt (1969); O'Neill (1981); Sargent (1972).

5. I have offered a graduate seminar where students work with a client drawn from the local community that has a business geographic-related need. The client was Bruce D. DeLaney, assistant vice president for administration–real estate, University of Florida Foundation. One of the properties in the large UF Foundation real estate holdings was being considered by a developer for a new 600 unit individual room-lease apartment complex. The following students assisted in the analysis reported on here, under my direction: Ed Borden, Justin Carasick, Evi De La Rosa-Ricciardi, Cindy Keyes, Jason Krejci, Dhiren Khona, Don Petrella, Barret Webber, and Craig Wells. This apartment market analysis is based upon Thrall (2001a).

Chapter 6

1. The discussion of background literature on office markets and the example of Tampa's office market presented here is a variation on the publications by Thrall and Amos (1999a, b, c, d). Paul Amos is director of GIS projects, Wharton School, University of Pennsylvania.

2. While the items might be viewed as common knowledge, this has not always been so. To the real estate analyst they may not appear as common knowledge. The pioneering research of the references associated with each item have been confirmed and elaborated. It was in this manner that each of the items became generally accepted in the industry for their role in determining office market rents. The references will assist the reader in learning more about the particular topic.

3. The terms *hedonic* and *heuristic* in this book, and generally elsewhere, refer to distinctly different concepts. *Heuristic*, as discussed in chapter 3, means "to tell a story" and in practice that often means using mathematical expressions. The term *hedonic* implies lack of formal structure. In practice, the term hedonic is reserved for regression equations as discussed in chapter 4. Myriad variables conjectured to capture some variation in the dependent variable are included in a regression equation, perhaps with the irrelevant variables being eliminated via a forward or backward stepwise regression.

4. Consider straight-line distance as the hypotenuse of a right triangle. The other two sides of the right triangle, easily calculated using Pythagoras's theorem, are referred to as *Manhattan distance*. Most cities in North America are layed out on a retangular grid, based on the Amer-

ican Survey System. So most major and minor roads go in a north–south, east–west direction. Therefore, other than straight north, south, east, or west, the traveler must proceed on the streets in a zigzag manner that approximates the Manhattan distance. Straight lines underestimates true distance that must be traveled. Manhattan distance closely approximates true distance that must be traveled in a city. For further discussion of how to measure geographic distances, and the implications of the measurement procedure chosen, see Thrall (1998), and Rodriguez et al. (1995).

5. Inventory data was provided by the Prince George's County Department of Economic Development.

6. The business establishment data they used were the *U.S. Establishment and Enterprise Microdata* file available from the U.S. Small Business Administration.

7. The nineteenth-century economist Alfred Marshall (1890, cited in Thrall 1991) believed that clustering of retail contributed to each individual store having a greater amount of sales. These spatial economic forces then contributed to the creation of nodes of retail within the city. Independently, geographers in the middle of the nineteenth century calibrated gravity models for trade area analysis using the concept of agglomeration.

8. Bartlesville, Oklahoma, is one of the few exceptions. Bartlesville is the home of Phillips Petroleum and several other closely related energy firms. The office market in Bartlesville depends on the needs of those few firms for employee office space, and to a lesser extent on the local economy that has grown because of the multiplier effect as discussed in chapter 2. The skyscrapers of Bartlesville have grown independent of the local urban land uses and are unique in the United States for a town so small. However, they were built in close proximity to oil wells.

9. Thrall et al. (1995) demonstrated this for St. Lucia County, Florida, using the legal property assessment file of the county. Time encompassed the start of the twentieth century through the early 1990s.

10. For a review of three-dimensional data and software, see Thrall (1999). For a review of office data that can be mapped as points, see Thrall (2000; see also Thrall 2001).

11. For a discussion of how to derive community preferences for amenities, which may also serve to be an amenity package that can attract and retain employees, see Thrall and McCartney (1991) and Thrall et al. (1988).

Chapter 7

1. Much of this chapter is based on publications by Thrall and del Valle (1996a, b, c; 1997a, b, c), Thrall et al. (1997; 1998a, b, c, d, e), and Thrall (1998b). Juan del Valle is Global Director of GIS for Blockbuster Entertainment. Gordon Hinzmann is Director of Strategic Development for Darden Restaurants (Red Lobster, Olive Garden, and Bahama Breeze).

2. The six-stage timeline presented here in figure 7.1 and table 7.3 is an extension of Thrall and del Valle's (1996) five-stage timeline and reflects advancements in desktop GIS; in turn, their time line was an extension of an earlier four-stage timeline presented by Goldstucker et al. (1978). Also see Eppli and Benjamin 1994.

3. Feng-shui has been practiced for thousands of years in China as a study of the natural environment and human-built environment and how these environments interact to affect the office, home, or building and their prosperity. The practice of feng-shui blends architecture, interior design, and the surrounding environment and knowledge of how people interact with buildings and the environment. The recommendations of feng-shui include how to modify the built and natural environment to improve luck, well being, and prosperity, and sites and situation most conducive for the same. In feng-shui, the prospective retail location is viewed holistically in the context of the natural and human-built environmental situation of the location.

4. *Signposting* is not a word that has a history of common application in location analysis. I am using it here to describe the concept of using features on the landscape as signposts that might mean, "based on experience this location will do well for the particular enterprise," or "this is a location to avoid."

5. Customer spotting techniques have further developed and have been applied by Ghosh and McLafferty (1987), Moloney (1989), and Rogers and Green (1978).

6. For a history of Starbucks and other coffee vendors in the United States, see Pendergrast (1999).

7. For a presentation of this technique using brand names of McDonalds, Arbys, Burger King, and Wendy's, traffic volume counts, and yellow pages on CD-ROM, see Thrall and Thrall (1993).

8. Data on branches of competing retail chains are more difficult to obtain. American Business Information (www.abii.com) sells corporate data, as does Dun and Bradstreet (*www.dnb.com*). However, these data are not necessarily available at the branch level, but may be available if the branches are franchises. American Business Information sells from a database of information on more than 10 million businesses in the United States. The database contains verified information on company names, addresses, SIC codes, telephone numbers, number of employees, sales volumes, and names of key decision makers. Dun and Bradstreet sells U.S. and international business information, such as credit reports and company financial information. Financial reports by franchise branch may be available in some instances from regional data vendors of credit information, such as www.business1.com. For further information on business geography data see Thrall (2001).

9. For a commentary on CACI's product, see Thrall (1998) and Thrall and McMullen (2000).

Chapter 8

1. For more discussion on the history of the hospitality industry, see PKF Consulting (1996).

2. There are no standard categorizations for the hospitality industry. An alternative to that presented here is PFK Consulting's (1996), whose market segmentation categories are based on price, function, location, market served, and distinctiveness of style or offerings (see Fox 1996).

3. For example, Stapleton Airport in Denver, Colorado, served as the sole traffic generator for many hotels and motels surrounding the airport. Until 1997 Stapleton was Denver's major commercial passenger airport, when the city of Denver replaced it with Denver International Airport, more than 30 miles away. Because of the change in traffic generator, the hospitality industry at Stapleton Airport does not have access to the highly focused market niche as before. No crossover niches appear to be available to substitute for the lost market. What had at one time been a locational advantage became a too narrowly defined market niche and ultimately a locational disadvantage.

4. This information is also useful in marketing the hospitality product. All major media categorize their viewers, listeners, and readers according to the same LSPs. The LSP of the target demographic population can then be matched with the appropriate media for advertising, thereby inducing demand.

5. Two- and three-stage least squares might be used to develop a more accurate, albeit more complex, model for projection. See Thrall et al. (1993) for an example.

6. The basic starting point for a hedonic model for R, NG, or OR is the same; namely, substitute NR or OR for R in equation. 8.1, and then proceed with forward or backward stepwise regression.

7. This procedure of expanding the regression coefficients to estimate their 'drift' across exogenous terms is from Casetti (1972).

8. Caution must be taken in constructing a hedonic model to avoid the use of identical terms. If two variables are identical, then they are perfectly correlated, voiding any results of the regression equation. For that reason, equation (8.5) has been specified without an intercept, as its addition to equation 8.4 would result in two column vectors of the value 1.0.

Chapter 9

1. The terms *mixed use* and MXD arise from the Urban Land Institute's 1976 publication, *Mixed-Use Developments: New Ways of Land Use* (Witherspoon et al. 1976). The book served to define those characteristics that are required to qualify a project as a true mixed-use development.

2. The background presentation on MXDs draws from the work of Schwanke (1987).

3. The concept and term *sense of place* is attributed to the significant work of geographer Ye Fu Tuan (1974).

References

Preface

Harvey, D. 1969. *Explanation in Geography* London: Edward Arnold.
Thrall, G. I. 1995. The stages of GIS reasoning *Geo Info Systems*. 5(2) 46–51.
Wofford, L. E. and G. I. Thrall. 1997. Real estate problem solving and geographic information systems: A stage model of reasoning. *Journal of Real Estate Literature* 5(2): 177–201

Chapter 1

Anikeeff, M. A., and G. R Mueller. 1998. Toward standardizing definitions by product type. In *Research Issues in Real Estate*, vol. 4 (M. A. Anikeeff and G. R. Mueller, eds.), pp. 89–108. Boston: Kluwer.
Berry, Brian J. L. 1991. *Long-Wave Rhythms in Economic Development and Political Behavior*. Baltimore, Md.: Johns Hopkins.
Brueggeman, W. B. and J. D. Fisher. 1996. *Real Estate Finance And Investments*. McGraw-Hill Higher Education.
Delisle, J. R. and J. Sa-Aadu (eds.). 1994. *Appraisal, Market Analysis, and Public Policy in Real Estate: Essays in Honor of James A. Graaskamp* (Real Estate Research Issues, Vol 1). New York: Kluwer Academic Publishers.
Doctrow, J. L., G. R. Mueller, L. L. Craig. 1999. Survival of the fittest: Competition, consolidation and growth in the assisted living industry. *The Journal of Real Estate Portfolio Management* 5(3): 225–234.
Fanning, S. F., T. V. Grissom, T. D. Pearson, T. Glisoon. 1995. *Market Analysis for Valuation Appraisals*. Chicago: Appraisal Institute.
Greer, Gaylon E., and M. D. Farrell. 1988. *Investment Analysts for Real Estate Decisions*, 2nd ed. Longman Financial Services Publishing.

Goetz, M. L., and L. E. Wofford. 1979. The motivation for zoning: Efficiency or wealth redistribution land. *Economics* 55(4): 472–485.

Golledge, R. G., and G. I. Thrall. 2000. The Fundamental Importance of Cognitive Mapping to Real Estate, paper presented at the Annual Meeting, American Real Estate Society, special session sponsored by the Homer Hoyt Institute, April 2000, Santa Barbara, CA.

Laposa, Steven, Grant Thrall, and David Watkins. 1997. The Development of a Senior Housing Expert Decision System—A Theoretical Model & Systems Framework. In *National Investment Conference Review (Research, Case Studies and Strategies for Investing in the Senior Living and Long Term Care Industries)*, vol. 5, pp. 3–16.

Marshall, Alfred. 1890. *Principles of Economics* (8th ed., 1946; London: Macmillan.)

Morrill, Richard, Gary Gaile, and Grant Ian Thrall. 1988. *Spatial Diffusion. Scientific Geography Series*, vol. 10. Newbury Park, Calif.: Sage Publications.

Mueller, G. R., and D. R. Wincott. 1995. Market analysis in the appraisal process. *The Appraisal Journal* 113(1): 86–94.

Mueller, G. R., and S. P. Laposa. 1998. The investment case for senior living and long-term care properties in an institutional real estate portfolio. In *Research Issues in Real Estate*, vol. 4 (M. A. Anikeeff and G. R. Mueller, eds.), pp. 171–183. Boston: Kluwer.

Mueller, G. R., and B. A. Ziering. 1992. Real estate diversification using economic diversification. *The Journal of Real Estate Research* 7(4):375–387.

Pyhrr, S. A., J. R. Cooper, P. D. Lapides. 1989. *Real Estate Investment: Strategy, Analysis, Decisions*. New York: John Wiley & Sons.

Pyhrr, S. A., Waldo L. Born, and James R. Webb. 1990. Development of a dynamic investment strategy under alternative inflation cycle scenarios. *Journal of Real Estate Research* 5(2): 177–193.

Thrall, G. I. 1987. *Land Use and Urban Form: The Consumption Theory of Land Rent*. London: Routledge/Methuen.

Thrall, G. I. 1991. The production theory of land rent. *Environment and Planning A* 23: 955–967.

Thrall, G. I., 2001. Data resources for real estate and business geography analysis. *Journal of Real Estate Literature* 9(2): 175–225.

Thrall, G. I. and S. McMullin. 2000. Trade area analysis: The buck starts here. *GeoSpatial Solutions* 10(6): 45–49.

Vernor, J. D. 1986. *Readings in Market Research for Real Estate: A Collection or Previously Published Articles*. New York: American Institute of Real Estate Appraisers.

Worzala, E. M., M. Lenk, and A. Silva. 1995. An exploration of neural networks and its application to real estate valuation. *Journal of Real Estate Research* 10(2): 185–201.

Wurtzebach, C. H., G. R. Mueller, and D. Machi. 1991. The impact of inflation and vacancy on real estate returns. *Journal of Real Estate Research* 6(2): 153–168.

Chapter 2

Anderson, J. E. 1985. The changing structure of a city: Temporal changes in cubic spline urban density patterns. *Journal of Regional Science* 25(4): 413–425.

Baen, J. S. 2000. The effects of technology on retail sales, commercial property values, and percentage rents. *Journal of Real Estate Portfolio Management*, 6(2): 185–201.

Baen, J. S., and R. S. Guttery. 1997. The coming downsizing of real estate: Implications of technology. *Journal of Real Estate Portfolio Management* 3(1): 1–18.

Bennett, R. J. 1980. *The Geography of Public Finance*. London: Methuen.

Berry, Brian J. L. 1964. Cities as systems. *Papers of the Regional Science Association* 13: 147–164.

Berry, Brian J. L. 1969. The factorial ecology of Calcutta. *American Journal of Sociology* 74, 445–491.

Berry, Brian J. L. 1970. *Geographic Perspectives on Urban Systems*. Englewood Cliffs, N.J.: Prentice Hall.

Berry, Brian J. L. 1991. *Long-Wave Rhythms in Economic Development and Political Behavior*. Baltimore, Md.: Johns Hopkins.

Berry, Brian J. L., Edgar Conkling, and D. Michael Ray. 1987. *Economic Geography*. Englewood Cliffs, N.J.: Prentice-Hall.

Born, W. L, and S. A. Pyhrr. 1994. Real estate valuation: The effect of market and property cycles. *Journal of Real Estate Research* 9(4): 455–486.

Brooks, David. 2000. *Bobos in Paradise: The New Upper Class and How They Got There*. New York: Simon & Schuster.

Bourassa, Steven C., and William G. Grigsby. 2000. Income tax concessions for owner-occupied housing. *Housing Policy Debate* 11(3): 521–546.

Buchanan, J. M. 1965. An economic theory of clubs. *Economica* 32: 1–14.

Casetti, Emilio. 1971. Equilibrium land values and population densities in an ideal.

Cheng, P., and R. T. Black. 1998. Geographic diversification and economic fundamentals in apartment markets: A demand perspective. *Journal of Real Estate Portfolio Management* 4(2): 93–105

Colby, C. C. 1933. Centrifugal and centripetal forces in urban geography. *Annals, Assoc. of American Geographers* 23: 1–20.

Demeny, P. 1977. The population of the underdeveloped countries. In *The Human Population (Scientific American*, Ed.), pp. 105–115. San Francisco: W. H. Freeman.

Eppli, M. J., and S. P. Laposa. 1997. A descriptive analysis of the retail real estate markets at the metropolitan level. *Journal of Real Estate Research* 14(3): 321–338.

Feagin, J. R. and R. Parker. 1990. *Building American Cities: The Urban Real Estate Game*, Englewood Cliffs, N.J.: Prentice Hall.

Freedman, R., and B. Berelson. 1977. The human population. In *The Human Population (Scientific American*, Ed.), pp. 3–25. San Francisco: W. H. Freeman.

George, H. 1879. *Progress and Poverty: San Francisco*. Reprinted by the Robert Schalkenbach Foundation, New York.

Grissom, T. V., K. Wang, and J. R. Webb. 1991. The spatial equilibrium of intra-regional rates of return and the implications for real estate portfolio diversification. *Journal of Real Estate Research* 7(1):59–71.

Hardin, Garrett. 1968. The tragedy of the commons. *Science*, 162: 1243–1248.

Hartshorn, Truman A. 1992. *Interpreting the City: An Urban Geography*. New York: John Wiley & Sons.

Hartzell, D. J., D. G. Shulman, and C. H. Wurtzebach. 1987. Refining the analysis of regional diversification for income-producing real estate. *Journal of Real Estate Research* 2(2): 85–95.

Hoyt, H. 1939. *The Structure and Growth of Residential Neighborhoods in American Cities*. Washington, D.C.: Federal Housing Administration.

Hoyt, H. 1966. *According to Hoyt-50 Years of Homer Hoyt (1916–1966)*. Washington, D.C.: The Homer Hoyt Institute.

Hoyt, H. 1933. *One Hundred Years of Land Values in Chicago*. Chicago: University of Chicago Press.

Kolbe, P. T., and R. D. Evans. 2001. Centripetal and Centrifugal Forces Impacting Industrial Real Estate In The Midsouth. Paper presented at the Annual Meetings, American Real Estate Society, April 2001, Santa Barbara, CA.

Kondratieff, N. D. 1935. The long wave in economic life. *Review of Economics and Statistics* 17: 1-5–115.

Miller, N. G. 2000. Retail leasing in a web enabled world. *Journal of Real Estate Portfolio Management* 6(2): 167–184.

Mueller, G. R. 1993. Refining economic diversification strategies for real estate portfolios. *Journal of Real Estate Research* 8(1): 55–68.

Mueller, G. R., and B. A. Zlering. 1992. Real estate portfolio diversification using economic diversification. *Journal of Real Estate Research* 7(4): 375–386.

Musgrave, R. A. 1959. *Theory of Public Finance*. New York: McGraw-Hill.

Pyhrr, S. A., W. L. Born, and J. R. Webb. 1990. Development of a dynamic investment strategy under alternative inflation cycle scenarios. *Journal of Real Estate Research* 5(2): 177–193.

Pyhrr, S. A., S. E. Roulac, and W. L. Born. 1999. Real estate cycles and their strategic implications for investors and portfolio managers in the global economy. *Journal of Real Estate Research* 18(1): 7–68.

Rabianski, J. S., and P. Cheng. 1997. Intrametropolitan spatial diversification. *Journal of Real Estate Portfolio Management* 3(2): 117–128.

Reid, C. E. 1977. Measuring residential decentralization of Blacks and Whites. *Urban Studies* 14: 353–357.

Roulac, S. E. 1994. Retail real estate in the 21st century: Information technology + time consciousness + unintelligent stores = intelligent shopping? NOT! *Journal of Real Estate Research* 9(1): 125–150.

Thrall, G. I. 1980. The consumption theory of land rent. *Urban Geography*, 1(4): 350–370.

Thrall, G. I. 1987. *Land Use and Urban Form*. London: Methuen.

Thrall, G. I. 1988. Statistical and theoretical issues in verifying the population density function. *Urban Geography* 9: 518–537.

Thrall, G. I. 1991. The production theory of land rent. *Environment and Planning A.*, 23: 955–967.

Thrall, G. I. 1992. Using the JOIN function to compare census tract populations between census years. *Geo Info Systems* 2(5): 78–81.

Thrall, G. I., Fandrich, and Thrall. 1995. The location quotient: Descriptive geography for the community reinvestment act. *Geo Info Systems* 5(6): 18–22.

Thrall, G. I. 1998. Common geographic errors in real estate analysis. *Journal of Real Estate Literature* 6(1): 45–54.

Thrall, G. I. 1999a. Real estate information technology: The future is today. In *GIS In Real Estate* (Gill Castle, ed.). Chicago: Appraisal Institute. 165–174.

Thrall, G. I. 1999b. iSite version 4.1. *Geo Info Systems*, 9(6): 47.

Thrall, G. I., and Shaun McMullin. 2000a. Getting to know your customer, CACI's Coder/Plus 3.00.18. *Geo Info Systems* 10(4): 42–43.

Thrall, G. I., and Shaun McMullin. 2000b. Trade area analysis: The buck starts here. *GeoSpatial Solutions* 10(6): 45–49.

Thrall, G. I., J. Casey and Aracibo Quintana. 2001. Trippin' on LSPs (Lifestyle Segmentation Profiles). *GeoSpatial Solutions* 11(4): 40–43.

Tiebout, C. 1956. A pure theory of local expenditures. *The Journal of Political Economy*, 64(5): 416–424.

Underhill, P. 1999. *Why We Buy: The Science of Shopping*. New York: Simon and Schuster.

Weicher, J. 2000. Comment on income tax concessions for owner-occupied housing. *Housing Policy Debate* 11(3): 547–560.

Wolf, P. 1981. *Land in America: Its Value, Use, and Control*. New York: Pantheon Books.

Yeates, M. 1997. *North American City*, 5th ed. New York: Addison-Wesley/Longman.

Chapter 3

Berry, Brian J. L, Edgar Conkling, and D. Michael Ray. 1987. *Economic Geography*. Englewood Cliffs, N.J.: Prentice-Hall.

Chisolm, Michael. 1965. *Rural Settlement and Land Use*. London: Hutchinson.

Dunn, Edgar S. 1953. *The Location of Agriculture Production*. Gainesville: University of Florida Press.

Hall, P. (ed.). 1966. *Von Thünen's Isolated State*. London: Pergamon.

Ricardo, David. 1817. *Principles of Political Economy and Taxation*. London.

Thrall, G. I. 1979. States of urban spatial structure: Implications of theoretical definitions upon urban computer simulations and mathematical models. *Environment and Planning A* 11: 23–35.

Thrall, G. I. 1980. The consumption theory of land rent. *Urban Geography* 1 (4):350–370.

Thrall, G. I. 1987. *Land Use and Urban Form*. London: Routledge/Methuen.

Thrall, G. I. 1991. The production theory of land rent. *Environment and Planning A* 23: 955–967.

von Thünen, Johann H. 1821. *Der isolierte Staat in Beziehung auf Landwirtschaft und Nationalekonomie*. Hamburg.

Wheeler, James O., Peter O. Muller, Grant Ian Thrall, and Timothy J. Fik. 1998. *Economic Geography*. New York: Wiley.

Chapter 4

Applebaum, W. 1965a. Can store location research be a science? *Economic Geography* 41:234–237.

Applebaum, W. 1965b. Measuring retail market penetration for a discount food supermarket—A case study. *Journal of Retailing* 41:1–47.

Applebaum, W. 1966. Methods for determining store trade areas, marketing penetration and potential sales. *Journal of Marketing Research*, 3:127–141.

Applebaum, W. 1968. *Guide to Store Location Research with Emphasis on Supermarkets*. Reading, Mass.: Addison-Wesley.

Applebaum, W., and S. B. Cohen. 1960. Evaluating store sites and determining store rents. *Economic Geography*, 36:1–35.

Batey, P., and P. Brown. 1995. From human ecology to customer targeting: the evolution of geodemographics. In *GIS for Business and Service Planning* (P. Longley and G. Clarke, eds.). New York: John Wiley & Sons. 77–103

Berry, Brian J. L. 1964. Cities as systems. *Papers of the Regional Science Association* 13:147–164.

Berry, Brian J. L. 1969. The factorial ecology of Calcutta. *American Journal of Sociology* 74: 445–491.

Berry, Brian J. L. 1970. *Geographic Perspectives on Urban Systems*. Englewood Cliffs, N.J.: Prentice Hall.

Berry, B.J.L., and W. L. Garrison. 1958. Functional bases of the central place hierarchy. *Economic Geography*, 34:145–154.

Berry, B.J.L., and J. B. Parr. 1988. *Market Centers and Retail Location: Theory and Applications*. Englewood Cliffs, N.J.: Prentice Hall.

Birkin, M. 1995. Customer targeting, geodemographics, and lifestyle approaches. In *GIS for Business and Service Planning* (P. Longley and G. Clarke, eds.), p. 104–149 New York: John Wiley & Sons.

Brown, L. A., M. A. Brown, and C. S. Craig. 1981. Innovation diffusion and entrepreneurial activity in a spatial context: Conceptual models and related case studies. In *Research in Marketing* (J. N. Sheth, ed.). Greenwich, Conn.: JAI Press. 69–115.

Buxton, T. T. 1993. Spatially-based models survive the test of time. *Business Geographics*, 1(2): 10–12.

Carey, H. C., 1858. *Principles of Social Science*. Philadelphia, Pa.: Lippincott.

Chatterjee, S., and B. Price. 1991. *Regression Analysis by Example*. New York: John Wiley & Sons.

Converse, RD. 1949. New laws of retail gravitation. *Journal of Marketing*, 14:379–384.

Cottrell, J. 1973. An environmental model of performance measurement in a chain of supermarkets. *Journal of Retailing*, 49(3):51–63.

Craig, C. S., A. Ghosh, and S. McLafferty. 1984. Models of retail location process: A review. *Journal of Retailing*, 60:5–36.

Davies, R. L. 1973. Evaluation of retail store attributes and sales performance. *European Journal of Marketing*, 7:89–102.

Davies, R. L. 1977. *Marketing Geography: With Special Reference to Retailing*. London: Methuen.

Davies, R. L., and D. S. Rogers. 1984. *Store Location and Store Assessment Research*. New York: John Wiley & Sons.

Donthu, N., et al. 1989. Estimating geographic customer densities using kernel density estimation. *Marketing Science* (Spring), 191–203.

Douglas, E. 1949. Measuring the general retail trading area—A case study: I. *Journal of Marketing* 13: 481–497.

Drummey, G. L. 1984. Traditional methods of sales forecasting. In *Store Location and Store Assessment Research* (R. L. Davies and D. S. Rogers, eds.). New York: John Wiley & Sons. 279–299

Epstein, B. J. 1978. Marketing geography: A chronicle of 45 years. *Proceedings of the Applied Geography Conference,* 1: 372–379.

Ferber, R. 1958. Variation in retail sales between cities. *Journal of Marketing,* 2:295–303.

Fotheringham, A. Stewart, and Morton E. O'Kelly. 1989. *Spatial Interaction Models: Formulations and Applications*. Dordrecht: Kluwer.

Ghosh, A., and S. McLafferty. 1987. *Location Strategies for Retail and Service Firms*. Lexington, Mass.: Lexington Books, D.C. Health and Company.

Goldstucker, J. L., D. N. Bellenger, T. J. Stanley, and R. L. Otte. 1978. *New Developments in Retail Trading Area Analysis and Site Selection*. Atlanta: College of Business Administration, Georgia State University.

Goodchild, Michael F. 2000. Spatial analysis and GIS. Pre-conference seminar presented at *2000* ESRI User Conference, Redlands, Calif.

Haynes, K., and A. S. Fotheringham. 1984. *Gravity and Spatial Interaction Models. Scientific Geography Series*, vol. 2. Thousand Oaks, Calif.: Sage Publications.

Heimsath, C. H. 1991. Small-area population estimation in absorption analysis. *Journal of Real Estate Research* 6(3): 315–326.

Hise, R. T., J. P. Kelly, M. Gable, and J. B. McDonald. 1983. Factors affecting the performance of individual chain store units: An empirical analysis. *Journal of Retailing*, 59(2):1–18.

Hoyt, H. 1958. *A Re-examination of the Shopping Center Market*. Technical Bulletin no. 33. Washington, D.C.: Urban Land Institute.

Hoyt, H. 1969. *The Location of Additional Retail Stores in the United States in the Last One-Third of the Twentieth Century, A Research Monograph*. New York: National Retail Merchants Association.

Huff, D. L. 1959. Geographical aspects of consumer behavior. *University of Washington Business Review* 18:27–37.

Huff, D. L. 1963. A probabilistic analysis of shopping center trade areas. *Land Economics* 39: 81–90.

Huff, D. L. 1964. Defining and estimating a trading area. *Journal of Marketing* 28:34–38.

Huff, D. L., and R. T. Rust. 1984. Measuring the congruence of market areas. *Journal of Marketing* 48: 68–74.

Ingene, C. A. 1984. Structural determinants of market potential. *Journal of Retailing* 60(1):37–64.

Ingene, C. A., and E. Yu. 1981. Determinants of retail sales in SMSAs. *Regional Science and Urban Economics* 11(4):529–547.

Jones, K. G., and D. R. Mock. 1984. Evaluating retail trading performance. In *Store Location and Store Assessment Research* (R. L. Davies and D. S. Rogers, eds.). New York: John Wiley & Sons. 333–360.

King, L. 1984. *Central place theory. Scientific Geography Series*, vol. 3. Thousand Oaks, Calif.: Sage Publications.

Lea, A. C. 1989. An overview of formal methods for retail site evaluation and sales forecasting: Part 1: *Operational Geographer* 7(2):8–17.

Lee, Y., and D. Koutsopoulos. 1976. A locational analysis of convenience food stores in metropolitan Denver. *Annals of Regional Science* 10(1):104–117.

Lord, J., and C. Lynds. 1981. The use of regression models in store location research: A review and case study. *Akron Business and Economic Review* 12(2):13–19.

Mahajan, V., A. K. Jain, and M. Bergier. 1977. Parameter estimation in marketing models in the presence of multicollinearity: An application of ridge regression. *Journal of Marketing Research* 14(4):586–591.

Moloney, T. 1989. A case study using a geographic information systems in food retailing. *Operational Geographer* 7:23–27.

Morrill, R., G. Gail, and G. I. Thrall. 1988. *Spatial Diffusion. Scientific Geography Series*, vol. 10. Thousand Oaks, Calif.: Sage Publications.

Moudon, A. V., and M. Hubner (eds.). 2000. Monitoring Land Supply with Geographic Information Systems: Theory, Practice, and Parcel Based Approaches. New York: John Wiley & Sons.

Odland, J. 1988. *Spatial Autocorrelation. Scientific Geography Series*, vol. 9. Thousand Oaks, Calif.: Sage Publications.

Okoruwa, A., H. O. Nourse, and J. V. Terza. 1994. Estimating sales for retail centers: An application of the Poisson gravity model. *Journal of Real Estate Research* 9(1):85–98.

Ravenstein, E. G. 1889. The laws of migration. *Journal of the Royal Statistical Society* 52:241–305.

Reilly, W. J. 1929. *Methods for the Study of Retail Relationships*. University of Texas Bulletin no. 2944. Austin: University of Texas.

Reilly W. J. 1931. *The Law of Retail Gravitation*. New York: Knickerbocker Press.

Ritchey, J. 1984. Developing a strategic planning data base. In *Store Location and Store Assessment Research* (R. L. Davies and D. S. Rogers, eds.). New York: John Wiley & Sons. 163–180

Rogers, D. S, and H. L. Green. 1978. A new perspective on forecasting store sales: Applying statistical models and techniques in the analog approach. *Geographical Review* 69(4):449–458.

Rust, R.A.B.J. 1986. Estimation and comparison of market area densities. *Journal of Retailing* 62(4):410–430.

Sabel, C. 1998. Peaks and troughs: Space-time cluster detection in rare diseases. Presented at the 1998 ESRI User's Conference, San Diego, Calif.

Simmons, M. 1984. Store assessment procedures. In *Store Location and Store Assessment Research* (R. L. Davies and D. S. Rogers, eds.). New York: John Wiley & Sons.

Stouffer, A. 1960. Intervening opportunities and competing migrants. *Journal of Regional Science* 1:1–20.

Thrall, G. I. 1986. Quantitative methods: Regression analysis. *Urban Geography* 7(6):547–555.

Thrall, G. I. 1988. Statistical and theoretical issues in verifying the population density function. *Urban Geography* 9(5):518–537.

Thrall, G. I. 1993. Using a GIS to rate the quality of property tax appraisal. *Geo Info Systems* 3:(3) 56–62.

Thrall, G. I. 1995. The stages of GIS reasoning. *Geo Info Systems* 5(2):46–51.

Thrall, G. 1998a. CACI Site Reporter with Scan/US. *Geo Info Systems* 8(6):46–49.

Thrall, G. 1998b. Common geographic errors of real estate Analysis. *Journal of Real Estate Literature*, 6(1):45–54.

Thrall, G. 1999a. Retail location analysis, step seven: Judgment. *Geo Info Systems* 9(2):36–37.

Thrall, G. I., and J. del Valle. 1997. The calculation of retail market areas: The Reilly model. *Geo Info Systems* 7(4): 46–49.

Thrall, G. I. 1999b. Zoning in on real estate data (Comps.Com and FW Dodge). *Geo Info Systems* 10(2):40–42.

Thrall, G. I. 2001. Data resources for real estate and business geography analysis. *Journal of Real Estate Literature* 9(2):175–225.

Thrall, G. I. 2000b. CEDDS (complete economic and demographic data source) from Woods & Poole Economics Inc. *Geo Info Systems*, 10(1), 36–37.

Thrall, G. I., and Paul Amos. 1999. Urban office market evaluation with GIS, Part 2: The decision to build. *Geo Info Systems* 9(6):40–46.

Thrall, G. I., and J. Casey. 2001. Deriving trade areas (MarketEdge, TrendMaps). *GeoSpatial Solutions* 11(11): 44–48.

Thrall, G. I., J. Casey, and Aracibo Quintana. 2001. Trippin' on LSPs (Experian's Mosaic and geoVue's iSITE). *GeoSpatial Solutions* 11(4): 40–43.

Thrall, G. I., and J. C. del Valle. 1996a. Retail location analysis: Antecedents. *Geo Info Systems* 6(6):48–52.

Thrall, G. I., and J. C. del Valle. 1996b. William Applebaum: Father of marketing geography. *Geo Info Systems* 6(9):50–54.

Thrall, G. I., and J. C. del Valle. 1996c. Calibrating an Applebaum analog market area model with regression analysis. *Geo Info Systems* 6(11):52–55.

Thrall, G. I., and J. C. del Valle. 1997a. The calculation of retail market areas: The Reilly model. *Geo Info Systems* 7(4):46–49.

Thrall, G. I., and J. C. del Valle. 1997b. Applied geography antecedents: Marketing geography. *Applied Geographic Studies* 1(3):207–214.

Thrall, G. I., and J. C. del Valle. 1997c. Antecedents of applied geography: Marketing geography. *Applied Geographic Studies* 1(3):207–214.

Thrall, G. I., J. del Valle, P. Amos, and K. McGurn. 1996. Measuring the development potential of an historic downtown. *Geo Info Systems* 6:44–49.

Thrall, G. I., J. del Valle, and G. Hinzmann. 1997. Retail location analysis with GIS: seven strategic steps. *Geo Info Systems* 7(11):42–46.

Thrall, G. I., J. del Valle, and G. Hinzmann. 1998a. Applying the seven-step site selection methodology to Red Lobster restaurants: Steps one and two. *Geo Info Systems* 8(2):40–43.

Thrall, G. I., J. del Valle, and G. Hinzmann. 1998b. Retail location analysis, step three: Assessing relative performance. *Geo Info Systems* 8(4):38–44.

Thrall, G. I., J. del Valle, and G. Hinzmann. 1998c. Retail location analysis, step four: identify situation targets. *Geo Info Systems* 8(6):38–43.

Thrall, G. I., J. del Valle, and G. Hinzmann. 1998d. Retail location analysis, step five: assess market penetration. *Geo Info Systems* 8(9):46–50.

Thrall, G. I., J. del Valle, and G. Hinzmann. 1998e. Retail location analysis, step 6: Identify markets for expansion. *Geo Info Systems* 8(9):42–45.

Thrall, G. I., M. McClanahan, and S. Elshaw Thrall. 1995. Ninety years of urban growth as described with GIS: A historic geography. *Geo Info Systems* 5(4):20–45.

Thrall, G. I., and S. McMullin. 2000a. Getting to know your customer, CACI's Coder/Plus 3.00.18. *Geo Info Systems* 10(4):42–43.

Thrall, G. I., and S. McMullin. 2000b. Trade area analysis: The buck starts here. *GeoSpatial Solutions* 10(6):45–49.

Thrall, G. I., and S. E. Thrall. 1991. Reducing investor risk: A GIS design for real estate analysis. *Geo Info Systems* 1(10):40–46.

Thrall, G. I., and S. E. Thrall. 1998. CensusCD+Maps (data review). *Geo Info Systems* 8(11): 47–49.

Wang, K., J. R. Webb, and S. Cannon. 1990. Estimating project-specific absorption. *Journal of Real Estate Research* 5(1):107–116.

Williams, Peter A., and A. Stewart Fotheringham. 1984. *The Calibration of Spatial Interaction Models by Maximum Likelihood Estimation with Program SIMODEL*. Geographic Monograph Series 7, Department of Geography, Indiana University, Bloomington.

Williamson, D., S. McLafferty, et al. 1998. Smoothing crime incident data: New methods for determining the bandwidth in kernel estimation. Presented at the 1998 ESRI User's Conference, San Diego, Calif.

Wilson, A. G. 1967. A statistical theory of spatial distribution models. *Transportation Research* 1:252–269.

Wilson, B. L. 1984. Modem methods of sales forecasting: Regression models. In *Store Location and Store Assessment Research* (R. L. Davies and D. S. Rogers, Eds.), pp. 301–308. New York: John Wiley & Sons.

Wofford, L, and E., and G. I. Thrall. 1997. Real estate problem solving and geographic information systems: A stage model of reasoning. *Journal of Real Estate Literature* 5(2):177–201

Zipf, G. K. 1949. *Human Behavior and the Principle of Least Effort*. Reading, Mass.: Addison-Wesley.

Chapter 5

Alves, W. R., and R. L. Morrill. 1975. Diffusion theory and planning. *Economic Geography 1975* 51: 290–304.

Barron, Lois M. 1998. The great gate debate. *Builder* 21 (2): 92–96.

Bourassa, Steven C., and William G. Grigsby. 2000. Income tax concessions for owner-occupied housing. *Housing Policy Debate* 11(3): 521–546.

Berry, B. J. L. 1972. Hierarchical diffusion: The basis of developmental filtering and spread in a system of growth centers. In *Growth Centers in Regional Economic Development* (N. Hanson, Ed.) New York: Free Press.

Blakely, Edward J., and Mary Gail Snyder. 1997. Places to hide. *American Demographics* 19(5): 22–25.

Bible, D. S., and C. H. Hsieh. 1996. Applications of geographic information systems for the analysis of apartment rents. *Journal of Real Estate Research* 12(1): 79–88

Bogdon, A., J. Follain, J. Goodman, D. Manson, and S. Brady. 1999. Research applications of the Multifamily Housing Institute's apartment database. *Journal of Real Estate Research* 7(2): 221–234.

Boyce, R. R. 1966. The edge of the metropolis; The wave theory analog approach. *British Columbia Geographical Series* 7:31–40.

Brown, L. A. 1975. The market and infrastructure context of adoption: A spatial perspective on the diffusion of innovation. *Economic Geography* 51:185–215.

Brown, L. A. 1981. *Innovation Diffusion: A New Perspective*. New York: Methuen.

Burgess, E. W. 1923. The growth of the city. *Proceedings of the American Sociological Society*, 18: 85–89.

Casetti, E. 1969. Why do diffusion processes conform to logistic trends? *Geographical Analysis* 1: 100–105.

Colenutt, R. J. 1969. Linear diffusion in an urban setting: An example. *Geographical Analysis* 1: 106–114.

Frew, J. R., G. Donald Jud, and Daniel T. Winkler. 1990. Atypicalities and apartment rent concessions. *Journal of Real Estate Research* 5(2): 195–201.

Hagerstrand, T. 1952. *On the propagation of innovation waves. Lund Studies in Geography B*, no. 44. Department of Geography. University of Lund, Sweden.

Hagerstrand, T. 1965. Aspects of the spatial structure of social communication and the diffusion of information. *Papers of the Regional Science Association* 16:27–42.

Hagerstrand, T. 1967a. *Innovation Diffusion as a Special Process* (Alan Pred, Trans.). Chicago: University of Chicago Press.

Hagerstrand, T. 1967b. On Monte Carlo simulation of diffusion. In *Quantitative Geography, Northwestern University Studies in Geography*, no. 13 (W. Garrison and D. Marble, eds.). Evanston, Ill.: Northwestern University.

Harvey, D. 1985. *Consolousness and the Urban Experience*. Baltimore, Md.: The Johns Hopkins University Press.

Harvey, J. 1992. *Urban Land Economics*. London: Macmillan.

Hoyt, H. 1939. *The Structure and Growth of Residential Neighborhoods in American Cities*. Washington, D.C.: Federal Housing Administration.

Miller, N. G., and M. A. Sklarz. 1986. Note on leading indicators of housing market price trends. *Journal of Real Estate Research* 1(1): 99–109.

Morrill, R. L. 1965. The Negro ghetto: Problems and alternatives. *Geographical Review* 55: 339–361.

Morrill, R. L. 1968. Waves of spatial diffusion. *Journal of Regional Science* 3: 1–17.

Morrill, R. L. 1970. The shape of diffusion in space and time. *Economic Geography* 46:259–68.

Morrill, R. L., G. L. Gaile, and G. I. Thrall. 1988. *Spatial Diffusion. Scientific Geography Series*, vol. 10. Newbury Park, Calif.: Sage Publications.

O'Neill, W. D. 1981. Estimation of a logistic growth and diffusion model describing neighborhood change. *Geographical Analysis*, 13: 391–397.

Pagliarl, J. L. and J. R. Webb. 1996. On setting apartment rental rates: A regression-based approach. *Journal of Real Estate Research* 12(1): 37–61.

Park, R. E., E. W. Burgess and R. D. McKenzie. 1925. *The City*. Chicago: University of Chicago Press. Reprinted in 1967.

Rostow, W. W. 1960. *Stages of Economic Growth: A Non Communist Manifesto*. New York: Cambridge University Press.

Sargent, C. S. 1972. Toward a dynamic model of urban morphology. *Economic Geography* 48: 357–74.

Sriram, V., and M. A. Anikeeff. 1991. Product-market strategies among development firms. *Journal of Real Estate Research* 7(1): 99–114.

Sumichraat, M., and M. Seldln. 1977. *Housing Market: The Complete Guide to Analysis and Strategy for Builders Lenders & Other Investors*. New York: Dow Jones-Irwin, Inc.

Thrall, G. I. 1983. The proportion of household income devoted to mortgage payments: A model with supporting evidence. *Annals of the Association of American Geographers* 73:220–230.

Thrall, G. I. 1987. *Land Use and Urban Form*. London: Routledge/Methuen.

Thrall, G. I. 1998. CACI Coder/Plus version 1, review. *Geo Info Systems* 8(6): 43–46.

Thrall, G. I. 2000. Zoning in on real estate data (Comps.Com and FW Dodge). *Geo Info Systems* 10(2): 40–42.

Thrall, G. I. 2001a. To build or not to build it? An apartment complex question. *GeoSpatial Solutions*, 11(3): 42–45.

Thrall, G. I. 2001b. Data resources for real estate and business geography analysis. *Journal of Real Estate Literature* 9(2): 175–225.

Thrall, G. I., and S. McMullin. 2000. Trade area analysis: The buck starts here. *GeoSpatial Solutions* 10(6): 45–49.

Thrall, G. I., M. McClanahan, and S. E. Thrall. 1995. Ninety years of urban growth as described with GIS: A historic geography. *Geo Info Systems* 5(4): 20–27.

Thrall, G. I., C. Sidman, S. E. Thrall, and Tim Fik. 1993. The cascade GIS diffusion model for measuring housing absorption by small area with a case study of St. Lucie County, Florida. *Journal of Real Estate Research* 8(3): 401–420.

Thrall, G. I., and S. E. Thrall. 1991. Reducing investor risk: A GIS design for real estate analysis. *Geo Info Systems* 1: 78–81.

Veblen, Thorstein Bunde. 1899. *The Theory of the Leisure Class: An Economic Study of Institutions.* Chicago.

Chapter 6

Archer, W. 1981. Determinants of location for general purpose office firms within medium size cities. *American Real Estate and Urban Economics Association Journal* 9(3):283–297.

Archer, W., M. Smith and D. Gatzlaff. 1990. *The Role of Visual Presence in Urban Office Location and Office Market Location.* Mimeograph. Gainesville: Department of Finance, Insurance, and Real Estate, University of Florida.

Black's Guide, Inc. 1997. *Black's Guide to Office Space-Tampa Bay/Southwest Florida.* Gaithersburg, Md.: Black's Guide, Inc.

Bollinger, C., K. Ihlanfeldt, and D. Bowes. 1997. *Spatial Variation in Office Rents within the Atlanta Region.* Atlanta: Policy Research Center and Department of Economics School of Policy Studies, Georgia State University.

Cannaday, R., and H. Kang. 1984. Estimation of market rent for office space. *The Real Estate Appraiser and Analyst* 50(6):67–72.

Carn, N., J. Rabianski, R. Racster, and M. Seldin. 1988. *Real Estate Market Analysis.* Englewood Cliffs, N.J.: Prentice Hall.

Caruthers, C. 1995. Occupancy rates improve sharply in downtown Tampa. *Tampa Bay Business Journal* 15 (27):1–2.

Clapp, J. 1980. The intrametropolitan location of office activities. *Journal of Regional Science* 20(3):387–399.

Clapp, J., H. Pollakowski, and L. Lynford. 1992. Intrametropolitan location and office market dynamics. *Journal of the American Real Estate and Urban Economics Association* 20(2): 229–257.

Frew, J. R., and G. D. Jud. 1988. The vacancy rate and rent levels in the commercial office market. *Journal of Real Estate Research* 3(1):1–8.

Garreau, J. 1991. *Edge City: Life on the New Frontier.* New York: Doubleday.

Gause, Jo Allen, et al. 1998. *Office Development Handbook*, 2nd ed. Washington, D.C.: Urban Land Institute.

Gad, G. 1985. Office location in Toronto. *Urban Geography* 6: 331–351.

Glascock, J., S. Jahanian, and C. Sirmans. 1990. An analysis of office market rents: Some empirical evidence. *Journal of the American Real Estate and Urban Economics Association* 18(1):105–119.

Greenberg, M. 1978. *Local Population and Employment Projection Techniques*. New Bruns-
wick, N.J.: Center for Urban Policy Research, Rutgers University.

Harris, C. and E. Ullman. 1945. The nature of cities. *Annals of the American Association of
Political and Social Sciences* 242:7–17.

Hough, D., and C. Kratz. 1983. Can "good" architecture meet the market test? *Journal of Urban
Economics* 15(1):40–54.

Howland, M., and D. Wessel. 1994. Projecting suburban office space demand: Alternative estimates of employment in offices. *The Journal of Real Estate Research* 9(3):369–383.

Kroll, C. 1984. *Employment Growth and Office Space along the 680 Corridor: Booming Supply
and Potential Demand in a Suburban Area*. Working Paper 84–75, Fisher Center for Real
Estate and Urban Economic Studies, University of California, Berkeley.

Marshall, Alfred. 1890. *Principles of Economics*. [8th ed., 1946; London: MacMillan.]

McDonald, J. 1993. Incidence of the property tax on commercial real estate: The case of
downtown Chicago. *National Tax Journal* 46(2):109–120.

McQueen, L. 1997. The office leasing pulse rate: Slow to sluggish. *Maddux Report* 14(6):57.

Mills, E. 1992. Office rent determination in the Chicago area. *Journal of the American Real
Estate and Urban Economics Association* 20(1):273–287.

Rodriguez, M., C. F. Sirmans, and A. P. Marks. 1995. Using geographic information systems
to improve real estate analysis. *Journal of Real Estate Research* 10(2):163–173.

Sarvis, M. 1989. What to look for in a new office facility. *The Journal of Business Strategy*
10(4):10–14.

Shilton, L. G., and J. R. Webb. 1991. Office employment growth and the changing function of
cities. *Journal of Real Estate Research* 7(1):73–90.

Sivitanidou, R. 1995. Urban spatial variation in office-commercial rents: The role of spatial
amenities and commercial zoning. *Journal of Urban Economics* 38(1):23–49.

Sivitanidou, R., and P. Sivitanides. 1995. The intrametropolitan distribution of R&D activities:
Theory and empirical evidence. *Journal of Regional Science* 35(3):391–415.

Terleckyj, N., and C. Coleman. 1997. *Regional Economic Growth in the United States: Projections for 1997–2025*. 1996 Regional Economic Projections. Washington, D.C.: NPA Data
Services, Inc.

Thompson B., and S. Tsolacos. 1999. Rent adjustment and forecasts in the industrial market.
Journal of Real Estate Research 17(2):151–167.

Thrall, G. I. 1991. The production theory of land rent. *Environment and Planning A*. 23(7):
955–967.

Thrall, G. I. 1998. Common geographic errors of real estate analysis. *Journal of Real Estate
Literature* 6(1):45–54.

Thrall, G. I. 2000. Zoning in on real estate data (Comps.Com and FW Dodge). *Geo Info Systems*
10(2):40–42.

Thrall, G. I. 2001. Real estate data and its categorization. *Journal of Real Estate Literature*, in
press.

Thrall, G. I. 1999. GIS software and data for three-dimensional visualization (ArcView 3.1,
ArcView 3D Analyst, MapFactory 3D Data). *Geo Info System* 9(2): 38–44.

Thrall, G. I., and P. Amos. 1999d. Urban office market evaluation with GIS, Part 4: Analyzing
the location decision. *Geo Info Systems* 9(11):44–49.

Thrall, G. I., and P. Amos. 1999c. Urban office market evaluation with GIS, Part 3: The decision
to locate. *Geo Info Systems* 9(9):39–45.

Thrall, G. I., and P. Amos. 1999b. Urban office market evaluation with GIS, Part 2: The decision
to build." *Geo Info Systems* 9(6):40–46.

Thrall, G. I., and P. Amos. 1999a. Urban office market evaluation with GIS, Part 1: Background
and setting. *Geo Info Systems* 9(4):321–36.

Thrall, G. I., and J. William McCartney. 1991. Using the Delphi method for GIS criteria. *Geo Info Systems* 1(1):46–52.

Thrall, G. I., M. McClanahan, and S. E. Thrall. 1995. Ninety years of urban growth described with GIS: A historical geography. *Geo Info Systems* 5(4):20–27.

Thrall, G. I., B. Swanson, and D. Nozzi. 1988. Greenspace acquisition and ranking program (GARP): a computer assisted decision strategy. *Computers, Environment, and Urban Systems* 12: 161–184.

Vandell, K., and J. Lane. 1989. The economics of architecture and urban design: Some preliminary findings. *Journal of the American Real Estate and Urban Economics Association* 17(2):235–260.

Warden, J. 1993. Industrial site selection: A GIS case study. *Geo Info Systems* 3(9):37–45.

Wheaton, W. 1984. The incidence of inter-jurisdictional differences in commercial property taxes. *National Tax Journal* 37(4):515–527.

Wheaton, William C., and Raymond G. Torto. 1994. Office rent indices and their behavior over time. *Journal of Urban Economics* 35(2):121–139.

Chapter 7

Applebaum, W. 1965a. Can store location research be a science? *Economic Geography*, 41; 234–237.

Applebaum, W. 1965b. Measuring retail market penetration for a discount food supermarket—A case study. *Journal of Retailing*, 41; 1–47.

Applebaum, W. 1966. Methods for determining store trade areas, marketing penetration and potential sales. *Journal of Marketing Research* 3: 127–141.

Applebaum, W. 1968. *Guide to Store Location Research with Emphasis on Supermarkets*. Reading, Mass.: Addison-Wesley.

Applebaum, W., and S. B. Cohen. 1960. Evaluating store sites and determining store rents. *Economic Geography* 36: 1–35.

Benjamin, J. D., G. Donald Jud, and D. T. Winkler. 1998. A simultaneous model and empirical test of the demand and supply of retail space. *Journal of Real Estate Research* 15(3): 297–307.

Berry, B.J.L., and W. L. Garrison. 1958. Functional bases of the central place hierarchy. *Economic Geography* 34; 145–154.

Brown, L. A., M. A. Brown, and C. S. Craig. 1981. Innovation diffusion and entrepreneurial activity in a spatial context: Conceptual models and related case studies. In *Research in Marketing* (J. N. Sheth, ed.), pp. 69–115. Greenwich, Conn.: JAI Press. 69–115.

Buxton, T. 1993. Spatially based models survive the test of time. *Business Geographics*, 1(2): 10–12.

Castle, Gilbert H. (ed.). 1998. *GIS in Real Estate: Integrating, Analyzing, and Presenting Locational Information* Chicago: Appraisal Institute.

Chatterjee, S., and B. Price. 1991. *Regression Analysis by Example*. New York: John Wiley & Sons.

Converse, R. D. 1949. New laws of retail gravitation. *Journal of Marketing*, 14: 379–384.

Cottrell, J. 1973. An environmental model of performance measurement in a chain of supermarkets. *Journal of Retailing*, 49(3): 51–63.

Craig, C. S., A. Ghosh, and S. McLafferty. 1984. Models of retail location process: A review. *Journal of Retailing*, 60; 5–36.

Davies, R. L. 1973. Evaluation of retail store attributes and sales performance. *European Journal of Marketing* 7: 89–102.

Davies, R. L. 1977. *Marketing Geography: With Special Reference to Retailing.* London: Methuen.

Davies, R. L., and D. S. Rogers. 1984. *Store Location and Store Assessment Research.* New York: John Wiley & Sons.

Douglas, E. 1949a. Measuring the general retail trading area—A case study: I. *Journal of Marketing*, 13: 481–497.

Douglas, E. 1949b. Measuring the general retail trading area—A case study: II. *Journal of Marketing,* 13: 46–60.

Drummey, G. L. 1984. Traditional methods of sales forecasting. In *Store Location and Store Assessment Research* (R. L. Davies and D. S. Rogers, eds.). New York: John Wiley & Sons.

Eppli, M. J. and J. D. Benjamin. 1994. The evolution of shopping center research: A review and analysis. *Journal of Real Estate Research* 9(1): 5–32.

Epstein, B. J. 1978. Marketing geography: A chronicle of 45 years. *Proceedings of the Applied Geography Conference,* 1: 372–379.

Ferber, R. 1958. Variation in retail sales between cities. *Journal of Marketing,* 22: 295–303.

Ghosh, A., and S. McLafferty. 1987. *Location Strategies for Retail and Service Firms.* Lexington, Mass.: Lexington Books, D.C. Health and Company.

Goldstucker, J. L., D. N. Bellenger, T. J. Stanley, and R. L. Otte. 1978. *New Developments in Retail Trading Area Analysis and Site Selection.* Atlanta: College of Business Administration, Georgia State University.

Hise, R. T., J. P. Kelly, M. Gable, and J. B. McDonald. 1983. Factors affecting the performance of individual chain store units: An empirical analysis. *Journal of Retailing,* 59(2): 1–18.

Hoyt, Homer. 1949. *Market Analysis of Shopping Centers.* Technical Bulletin No. 12. Washington, D.C.: The Urban Land Institute.

Hoyt, Homer. 1958. *A Re-examination of the Shopping Center Market.* Technical Bulletin no. 33. Washington, D.C.: The Urban Land Institute.

Hoyt, Homer. 1969. *The Location of Additional Retail Stores in the United States in the Last One-Third of the Twentieth Century, A Research Monograph.* New York: National Retail Merchants Association.

Ingene, C. A. 1984. Structural determinants of market potential. *Journal of Retailing,* 60(1); one 37–64.

Ingene, C. A., and E. Yu. 1981. Determinants of retail sales in SMSAs. *Regional Science and Urban Economics* 11(4): 529–547.

Jones, K. G., and D. R. Mock. 1984. Evaluating retail trading performance. In *Store Location and Store Assessment Research* (R. L. Davies and D. S. Rogers, eds.). New York: John Wiley & Sons.

King, L. 1984. *Central place theory. Scientific Geography Series,* vol. 3. Thousand Oaks, Calif.: Sage Publications.

Lea, A. C. 1989. An overview of formal methods for retail site evaluation and sales forecasting: Part 1. *Operational Geographer,* 7(2); 8–17.

Lee, Y., and D. Koutsopoulos. 1976. A locational analysis of convenience food stores in metropolitan Denver. *Annals of Regional Science* 10(1); 104–117.

Longley, Paul, and Graham Clarke (eds.). 1995. *GIS For Business and Service Planning.* London: Geo Information International.

Moloney, T. 1989. A case study using a geographic information system in food retailing. *Operational Geographer,* 7, 23–27.

Morrill, R., G. Gail, and G. I. Thrall. 1988. *Spatial diffusion. Scientific Geography Series,* vol. 10, Thousand Oaks, Calif.: Sage Publications.

Odland, J. 1988. *Spatial autocorrelation. Scientific Geography Series,* vol. 9, Thousand Oaks, Calif.: Sage Publications.

Pendergrast, M. 2000. *Uncommon Grounds: The History of Coffee and How It Transformed Our World*. New York: Basic Books.

Rogers, D. S., and H. L. Green. 1978. A new perspective on forecasting store sales: Applying statistical models and techniques in the analog approach. *Geographical Review*, 69: 449–458.

Sharkawy, M. A., X. Chen, and F. Pretorius. 1995. Spatial Trends of Urban Development in China *Journal of Real Estate Literature* 3(1): 47–59.

Simmons, M. 1984. Store assessment procedures. In *Store Location and Store Assessment Research* (R. L. Davies and D. S. Rogers, Eds.), pp. 263–278. New York: John Wiley & Sons.

Thrall, G. I. 1979. A geographic criterion for identifying property tax assessment inequity. *Professional Geographer* 31(3): 278–283.

Thrall, G. I. 1998a. Common geographic errors of real estate analysis. *Journal of Real Estate Literature*, 6(1): 45–54.

Thrall, G. I. 1998b. Retail location analysis, step seven: Judgment, implementation, monitoring. *Geo Info Systems*, 8(11). 36–37

Thrall, G. I. 1998c. CACI Coder/Plus. *Geo Info Systems* 8(6):43–46.

Thrall, G. I. 2001. Data resources for real estate and business geography analysis. *Journal of Real Estate Literature* 9(2): 175–225.

Thrall, G. I., and J. C. del Valle. 1996a. Retail location analysis: Antecedents. *Geo Info Systems*, 6(6); 48–52.

Thrall, G. I., and J. C. del Valle. 1996b. William Applebaum: Father of marketing geography. *Geo Info Systems, 6(9)*; 50–54.

Thrall, G. I., and J. C. del Valle. 1996c. Calibrating an Applebaum analog market area model with regression analysis. *Geo Info Systems*, 6(11); 52–55.

Thrall, G. I., and J. C. del Valle. 1997a. The calculation of retail market areas: The Reilly model. *Geo Info Systems* 7(4); 46–49.

Thrall, G. I., and J. C. del Valle. 1997b. Applied geography antecedents: Marketing geography. *Applied Geographic Studies*, 1(3); 207–214.

Thrall, G. I., and J. C. del Valle. 1997c. Antecedents of applied geography: Marketing geography. *Applied Geographic Studies*, 1(3): 207–214.

Thrall, G. I., J. C. del Valle, P. Amos, and K. McGurn. 1996. Measuring the development potential of an historic downtown. *Geo Info Systems* 6(4); 44–49.

Thrall, G. I., J. del Valle, and G. Hinzmann. 1997. Retail location analysis with GIS: Seven strategic steps. *Geo Info Systems* 7(11); 42–46.

Thrall, G. I., J. del Valle, and G. Hinzmann. 1998a. Applying the seven-step site selection methodology to Red Lobster restaurants: Steps one and two. *Geo Info Systems* 8(2); 40–43.

Thrall, G. I., J. del Valle, and G. Hinzmann. 1998b. Retail location analysis, step three: Assessing relative performance. *Geo Info Systems* 8(4): 38–44.

Thrall, G. I., J. del Valle, and G. Hinzmann. 1998c. Retail location analysis, step four, identify situation targets. *Geo Info Systems* 8(6); 38–43.

Thrall, G. I., J. del Valle, and G. Hinzmann. 1998d. Retail location analysis, step five, assess market penetration. *Geo Info Systems* 8(9): 46–50.

Thrall, G. I., J. del Valle, and G. Hinzmann. 1998e. Retail location analysis, step 6: Identify markets for expansion. *Geo Info Systems* 8(9); 42–45.

Thrall, G. I., M. McClanahan, and S. Elshaw Thrall. 1995. Ninety years of urban growth as described with GIS: A historic geography. *Geo Info Systems*, 5(4); 20–45.

Thrall, G. I., and Shaun McMullin. 2000. Getting to know your customer; CACI Coder/plus software and data. *Geo Info Systems* 10(4): 42–43.

Thrall, G. I., and S. E. Thrall. 1993. Commercial data for the business GIS (Part one). *Geo Info Systems*, 3(7):63–68.

Urban Land Institute. 1999. *Shopping Center Handbook*. Washington, D.C.: ULI.

Wilson, B. L. 1984. Modern methods of sales forecasting: Regression models. In *Store Location and Store Assessment Research* (R. L. Davies and D. S. Rogers, Eds.),pp. 301–318 New York: John Wiley & Sons.

Chapter 8

Casetti, E. 1972. Generating models by the expansion method: Applications to geographic research. *Geographical Analysis* 4:81–91.

Fox, John A. 1996. Products and their markets. In *Hotel Development* (PFK Consulting, Ed.), pp. 21–28. Washington, D.C.: Urban Land Institute.

PFK Consulting. 1996. *Hotel Development*. Washington, D.C.: Urban Land Institute.

Schmitt, Genevieve. 1999. A Rocky Mountain getaway. *Asphalt Angels*, May/June, pp. 10–11.

Thrall, G. I. 2001. Data resources for real estate and business geography analysis (6). *Journal of Real Estate Literature* 9(2): 176–225.

Thrall, G. I., C. Sidman, S. E. Thrall, and T. Fik. 1993. The cascade GIS diffusion model for measuring housing absorption by small area with a case study of St. Lucie County, Florida. *Journal of Real Estate Research* 8(3):401–420.

Thrall, G. I., and S. E. Thrall. 1993. Commercial data for the business GIS. *Geo Info Systems* 3(7):63–68.

Wallis, Michael. 1981. *Route 66: The Mother Road*. New York: St. Martin's Griffin.

Winfree, W. Martin Jr. 1996. A historical perspective. In *Hotel Development* (PFK Consulting, ed.), pp. 1–10. Washington, D.C.: Urban Land Institute.

Wirth, Linda Schrier. 1996. Market segmentation and analysis. In *Hotel Development* (PFK Consulting, ed.), pp. 11–20. Washington, D.C.: Urban Land Institute.

Chapter 9

Schwanke, Dean. 1987. *Mixed-Use Development Handbook*. Washington, D.C.: Urban Land Institute.

Thrall, G. I. 2001. Data resources for real estate and business geography analysis. *Journal of Real Estate Literature* 9(2): 175–225.

Thrall, G. I., M. McClanahan, and S. E. Thrall. 1995. Ninety years of urban growth as described with GIS: A historic geography. *Geo Info Systems*, 5(4), 20–27.

Thrall, G. I., J. del Valle, P. Amos, and K. McGurn. 1996. Measuring the development potential of an historic downtown. *Geo Info Systems*, 6(4), 44–49.

Tuan, Ye Fu. 1974. *Topophilia, A Study of Environmental Perception, Attitudes, and Values*. New York: Columbia University Press.

Witherspoon, Robert, Jon P. Abbett, and Robert Gladstone. 1976. *Mixed-Use Developments: New Ways of Land Use*. Washington, D.C.: Urban Land Institute.

Chapter 10

Clapp, J. M., M., M. Rodriguez, G. I. Thrall. 1997. How GIS can put urban economic analysis on the map. *Journal of Housing Economics*. 6(4): 368–386.

Thrall, G. I. 2001a. Data resources for real estate and business geography analysis. *Journal of Real Estate Literature* 9(2): 175–225.

Thrall, G. I. 2001b. Business geography data resources. Part 1: Private data vendors. *GeoSpatial Solutions* 11(5): 42–45.

Thrall, G. I. 2001c. Business geography data resources: Part 2: Federal government data. *GeoSpatial Solutions* 11(6): 42–49.

Index